# Dinner with Darwin

# Dinner *with* Darwin

FOOD, DRINK,
AND EVOLUTION

*Jonathan Silvertown*

The University of Chicago Press
Chicago and London

The University of Chicago Press, Chicago 60637
The University of Chicago Press, Ltd., London

Published 2017
Paperback edition 2020
Printed in the United States of America

29  28  27  26  25  24  23  22  21  20       1  2  3  4  5

ISBN-13: 978-0-226-24539-3 (cloth)
ISBN-13: 978-0-226-76009-4 (paper)
ISBN-13: 978-0-226-48923-0 (e-book)
DOI: https://doi.org/10.7208/chicago/9780226489230.001.0001

Library of Congress Cataloging-in-Publication Data

Names: Silvertown, Jonathan W., author.
Title: Dinner with Darwin: food, drink, and evolution /
    Jonathan Silvertown.
Description: Chicago; London: The University of Chicago
    Press, 2017. | Includes bibliographical references.
Identifiers: LCCN 2017007291 | ISBN 9780226245393 (cloth:
    alk. paper) | ISBN 9780226489230 (e-book)
Subjects: LCSH: Food habits. | Diet. | Evolution—Social
    aspects. | Dinners and dining.
Classification: LCC GT2855 .S58 2017 | DDC 394.1/2—dc23
    LC record available at https://lccn.loc.gov/2017007291

♾ This paper meets the requirements of
ANSI/NISO Z39.48–1992 (Permanence of Paper).

For my brother, Adrian

# Contents

# 1

# *An Invitation to Dinner*

There are too many books on food. That is a contrary, self-defeating statement to read in yet another book on food, but have you never wondered what there could possibly be left to say on the subject? This thought certainly occurred to me one afternoon as, careful not to awaken the exhausted students dozing in the window niches, I browsed the food section of the well-stocked library at University of California, Davis. There, every facet of food and drink from Artichoke to Zinfandel is researched and taught. Merely scanning the titles on the shelves was an education in itself. *The Complete Idiot's Guide to Smoking Foods* has presumably prevented many of its intellectually challenged readers from mistaking barbecue for pipe tobacco.

Who would have thought that a large volume on *Bubbles in Food* would require an even more voluminous follow-up: *Bubbles in Food 2*? Or that shelved among books on meat and pies, a book called *A Diet of Tripe* would not contain instructions on how to survive on the cooked stomach lining of cows, but a diatribe against food faddism in general and vegetarianism in particular. Across the aisle was *No More Bull!*, a vegan manifesto written by an erstwhile cowboy. If the authors of these two books ever met, I'd like to think that the author of *Handheld Pies* would also be there to supply the ammo. On a more serious note (well, almost), the proceedings of an Oxford Symposium on food and cookery yielded a mother lode of erudition on "Ancient Jewish Sausages,"

"Transylvanian Charcoal-Coated Bread," "Shad Planking," and UFOs ("Unidentified Fermented Objects"). For the more industrially minded cook, there was the book *Food Processing by Ultra High Pressure Twin-Screw Extrusion*.

So, just in case there really is a surfeit of books on food, I want us to pretend that what you now hold in your hands is not so much a book as an invitation to dinner in the hope that, if you are like me, you can never have too many of those. This, though, is going to be a dinner with a difference. This is going to be a dinner of the mind. Of course, every meal belongs in the brain, for this is where the sensations created by eating food are processed and perceived, but my invitation is to think about what we eat in a different way.

For example, what do eggs, milk, and flour have in common? If you enjoy cooking, you will immediately recognize that these are the chief ingredients of pancakes, but there is another much more interesting answer too. Eggs, milk, and seeds (from which flour is made) are each designed by evolution to nourish offspring. Ponder this simple fact deeply, and a whole story will explode from the idea. This book tells that story, not just for pancake ingredients, but for a fourteen-chapter meal.

Everything we eat has an evolutionary history. Every supermarket shelf is packed with the products of evolution, though the label on the poultry will not remind us of this with a Jurassic sell-by date, nor will the tickets in the produce aisle betray the fact that corn has a 6,000-year history of artificial selection by pre-Columbian Americans. Any shopping list, each recipe, every menu, and all ingredients contain a silent invitation to dine with the father of evolutionary understanding, Charles Darwin.

Until Darwin's book *On the Origin of Species* was published in 1859, the obvious presence of design in nature—like the perfect nutritive qualities of milk as food for babies—was held to be self-evident proof of the existence of a designer and that designer must be God. But Darwin came up with another answer: natural selection. Everything in nature varies, and a proportion of that variation is usually inherited. Adults vary in their tolerance of milk, for example, and this tolerance is largely genetic. Natural selection is the winnowing of inherited variation that, little by little, generation by generation, cumulatively improves the functioning of organisms, as the genetic variants that are better suited to local conditions multiply at the expense of the less well-adapted ones. This process of gradual evolution is blind and free from any intention, plan, or goal.

Evolution by natural selection produces design without any designer. Contradictory though this may sound, it's the process that not only produced our food, but also produced us. Our relationships with food demonstrate evolution in ourselves and in what we eat. Learning about these relationships can nourish the mind as well as the stomach. If you have a taste for long words, you might call it "evolutionary gastronomy," or you could just say that we are going to make a meal out of evolution.

The very first chapter of *On the Origin of Species* is about the domestication of plants and animals because Darwin realized that the process of artificial selection that breeders use to produce new varieties is analogous to natural selection. The enormous, cumulative changes wrought by breeders demonstrate what can also be achieved by a gradual process of natural selection. At first sight, it may seem strange that plants and animals are so malleable that we can divert them from their own evolutionary path and shape them so readily to meet our particular needs. The reason that this is possible is that artificial selection is itself an evolutionary process, so that rather than working against evolution, we are in fact working with it.

Artificial selection directs the evolution of plants and animals in the same way that an engineer directs the flow of a river by shaping the landscape with canals, dams, and levees, enabling gravity to channel water in the desired direction. Breeders direct the flow of genes by selecting which individuals will produce the next generation, and genetics does the rest. Two things are needed for this to work: there must be variation among individuals in the characteristics the breeder wants to influence, and a proportion of this variation must be inherited (heritable).

It was evolution by natural selection that equipped eggs, milk, and seeds with the properties that allow us to turn them into pancakes. To discover how this happened, let's start with an egg, that metaphor for beginnings and very possibly the most versatile food that evolution has given us. Eggs are not only delicious fried, boiled, scrambled, poached, or even pickled, but as ingredients they have almost magical powers, levitating soufflés, cakes, quiche, and meringues, and stabilizing the otherwise immiscible oil- and water-based components of mayonnaise and sauces. Eggs are so nutritious because they contain all the food needed for the development of a chick, and they keep so well in the kitchen because the shell is designed by evolution to prevent the egg from drying out and to help protect the contents from the bacteria and fungi that cause rot and decay. How did these useful properties of eggs evolve?

Chickens make eggs and eggs make chickens, hence the phrase "Chicken-and-Egg," which uses the life cycle of the fowl as a metaphor for any circular situation that has no discernible starting point. But if we take an evolutionary perspective, the chicken-and-egg conundrum is easily unscrambled: eggs evolved before chickens. Birds are the modern descendants of a lineage of reptiles that included the iconic dinosaur predator *Tyrannosaurus rex*. We now know from wonderfully preserved fossils found in China that many dinosaurs were feathered. So, the chicken's feathers are inherited from reptilian forebears, just like the hen's egg. In fact, dinosaurs also nested, and it seems that males as well as females brooded their eggs, just as some birds do. Birds really are dinosaurs.

Some of the first fossil dinosaur eggs to be scientifically described were found in 1859, the very year that Darwin published *On the Origin of Species*. They were discovered in Provence, southern France, by Father Jean-Jacques Pouech, a Catholic priest and naturalist who quite reasonably believed they must have belonged to a giant bird. It seems somehow fitting that the country that gave us the omelet and the soufflé is also where the modern egg's reptilian forebears were first discovered. Although dinosaur eggs have now been found all over the world, southern France is still a global hotspot for these fossils.

In the evolutionary history of life, an egg protected by a mineral shell was a reptilian invention, but beneath the shell is something of even greater antiquity that was a game-changer for life on land. The first animals to make the transition from ocean to land were amphibians but, like their modern representatives such as salamanders and frogs, their jelly-like eggs lacked protection from drying out in air. So, although the adults could survive on dry land, their eggs still had to be laid in water or they would shrivel and die.

The game-changer was the evolution of a membrane called the amnion, which encloses the embryo in a bag of fluid called the amniotic sac. The amniotic sac is typical of the way that evolution solves problems by the most available route. You can almost hear the salesman's cry as it echoes from the primeval swamp forests of the Upper Carboniferous, 310 million years ago: "Embryo drying out? Here's a new idea! Pop it in this bag of pond water." In fact, pancakes contain a second example of this very adaptation to life on land.

The evolutionary origin of seeds 360 million years ago is a remarkably similar story to the origin of the amnion, that essential step on the road to the hen's egg. Just as the amniotic sac was the animal solution to the problem of how to reproduce on land, so the seed was the plant solution to the same problem. The first seed plants evolved from terrestrial ancestors that required liquid water in a wet environment for egg and sperm to meet, just as modern ferns and mosses do today. Seed plants are to ferns as amniotes are to amphibians. In both cases, it was the evolution of a liquid-filled bag to hold the embryo and then the addition of a desiccation-resistant package around it containing plenty of food that was the big innovation.

And so we come to the evolutionary story of the third pancake ingredient, milk. Feeding our young on milk is a defining characteristic of us mammals, and all species do it in the same way from glands that are specialized for lactation. The clue is in the name, for mammals are animals with mammaries and produce milk by the bucket. And what a bucket! The average milk cow in the United States produces $9\frac{1}{2}$ tons of milk a year. The largest mammal is the blue whale. It is estimated that a nursing female weighing 100 tons produces nearly 500 pounds of milk a day for her calf, containing enough energy to sustain 400 people daily.

The evolutionary history of mammals, birds, plants, and life itself was known in only the sketchiest detail in Darwin's day, but it is now being revealed in greater and greater detail at a breath-taking pace. This is due to the ease with which we can now read and compare the genomes of different species. A genome is essentially a recipe book that contains all the instructions needed for the machinery of a cell to turn, for example, a fertilized egg into a chicken and for that chicken's cells and organs to do all the things that chickens do, including, and most importantly for both evolution and cuisine, make more chickens!

The genome is written in a chemical alphabet made of the building blocks of nucleic acids. There are only four letters (different nucleic acid building blocks) in this alphabet, but the combination of these letters into a DNA sequence can spell out very long and complicated recipes for cells to make proteins of all kinds. These recipes are in fact genes. Some of the proteins made by gene recipes—for example, in egg yolk—are food molecules. Other genes make a special class of proteins called enzymes. These speed up (catalyze) biochemical reactions such as the breakdown of starch into simple sugars by the enzyme amylase that is present in our saliva. Yet another class of gene is a switch that turns other genes on or off. The cell is like a tiny automated kitchen cooking up tens of thousands of recipes at any one time and constantly tweaking the output from these recipes up or down according to need.

Genomes contain not just active genes, but also pseudogenes, which are the ghosts of genes past. These are the recipes that are not used anymore, though they are still copied from one edition of the recipe book to the next every time a new generation is born. Functioning genes are faithfully copied and corrected. Any fatal errors that happen to arise are removed by natural selection by their carriers dying before they can pass on their genetic flaws to any offspring. However, once a gene no longer functions, copying errors do not affect processes essential to survival or reproduction, and so errors accumulate, causing the gene sequence to become more and more nonsensical with time. The longer ago a pseudogene has lost its function, the more its sequence will differ from the sequence in genes that still function. So, after a few hundred generations of disuse, a recipe that began as "Beat the white of one egg" might become "Eat the white of one egg" and after a few thousand generations, "Tea the wheti of done gg."

The sequences of the different genes involved in manufacturing egg yolk and milk reflect the evolutionary transition that took place from egg-laying ancestors to live-bearing mammals that nurture their young with milk. In our own mammalian lineage, yolk genes like those found in chickens became pseudogenes 30–70 million years ago. This occurred long after the appearance of the genes that make milk proteins, so there must have been an intermediate stage when mammals laid eggs and also produced milk. A comparison of the genomes of the chicken and the platypus, which is an egg-laying mammal, discovered that one of the genes that makes the protein in egg yolk in chickens is still found in a functional state in the platypus. So, as one would expect, the platypus genome contains milk protein as well as yolk protein genes, testifying to the species' status as a relict of the transition that took place in mammals from egg-laying to live-bearing.

Eggs, seeds, and milk were all a solution to a fundamental question familiar to every parent: How do we protect and nurture baby? Surreal though it sounds, the evolution of each of these three pancake ingredients was a turning point in the evolution of life on Earth.

A pancake is not normally served as an appetizer, but I hope that this one has created a sense of anticipation for what is to come. Now, let me guide you through the rest of the menu. All ingredients are guaranteed fresh and locally sourced. The knowledge suppliers are exhaustively listed at the end of the book. As an aside, let me mention that you can follow the plan that I have set out, or you can dine *à la carte* if you wish, making your own selection of courses in any order. Some items you won't find on the menu are coffee, fruit, and nuts because these were served in my previous book, *An Orchard Invisible: A Natural History of Seeds*. Don't you hate it when foods repeat?

Cooking is fundamental to human nutrition, and, as we shall discover in chapter 2, it is a truly ancient practice that was pivotal in the evolution of humans. So, too, was the consumption of shellfish that sustained small bands of our species as they migrated out of Africa some 70,000 years ago (chapter 3). Agriculture, founded upon the domestication of plants and animals, is the basis of our diet today. Like the plaited dough in a challah loaf, chapter 4 entwines the story of crop domestication at the dawn of agriculture with the history of bread.

The following two chapters are about how we evolved the senses of taste and smell that enable us to respond to the chemistry of plants and other food. This is how we are able to make life-preserving choices about what is edible and what is not. These topics are served with soup (chapter 5) and fish (chapter 6).

We have set the course of evolution in our crops, but in consuming them they have shaped our evolution too. But beware, whatever the groaning shelf-loads of Paleo diet books might try to tell you, evolution is not destiny. We are not better off eating mammoth amounts of meat because this is how evolution shaped us in the Paleolithic (chapter 7). We are omnivores, and evolution does not dictate how we must behave or what we must eat beyond some very obvious constraints. "Never eat anything bigger than your head" has always seemed like sound advice to me. And, as the food writer Michael Pollan has said, three simple rules that you already know contain the best health advice you can get: Eat food, mainly plants, and not too much.

Just how little evolution constrains our diet can be readily demonstrated through the vegetables that we eat (chapter 8). We have found ingenious ways to process even unpromising and poisonous plants into delicious foodstuffs and can consequently eat more than 4,000 species. If you wish to celebrate the diversity of plants that we are able to eat, you can emulate members of the Botanical Society of Scotland, who in 2013 held a competition for the Christmas cake recipe with the most species of plant among its ingredients. The winning recipe was baked and contained 127 species belonging to 54 plant families. The topping alone included candied pecans, walnuts, cashews, almonds, pine kernels, sesame seeds, angelica, coconut flakes, and chocolate-covered coffee beans, and was decorated with the dried and sugar-dusted flowers of violets, primroses, lavender, rosemary, borage, winter jasmine, daisy, and calendula.

Plants cannot run or fly away from their enemies as animals can, and so evolution forces them to adopt a defensive strategy instead. Like a nerdy kid in school who has no athletic ability, plants compensate for being slow and vulnerable in the field by excelling in the chemistry lab. Thus, the simple fact that plants can't flee has profound consequences for cuisine. As we discover in chapter 9, this is responsible for the flavor of spices; the bite of mustard and horseradish; the fiery piquancy of ginger and chili; and all the medicinal effects of plants into the bargain.

In chapter 10 we entertain some culinary indulgence in the form of desserts that pander to our primitive desires for sugar and fat. By chapter 11, the cheese I have prepared for you has reached a satisfying ripeness with an aroma that demands attention. Unlike anything else we eat, cheese has no direct equivalent in nature but this confection of milk and microbes contains an evolutionary ferment. And talking of fermentation, in chapter 12 we take to the bottle like a fruit fly takes wing to rotting fruit. Oenophile and fly are both attracted by alcohol, for which we are indebted to yeast and its own long evolutionary relationship with demon drink.

The penultimate chapter (chapter 13) reflects on a question that is so fundamental to dining that it is always taken for granted. The question is "Why do we share food?" The evolutionary answers should make excellent conversation at any mealtime. The conclusion is that even restaurants have an evolutionary origin. Finally, in chapter 14, we look at the future of food and the controversial role that genetic modification will play in its evolution. Now, please follow me to the table and *Bon appétit!*

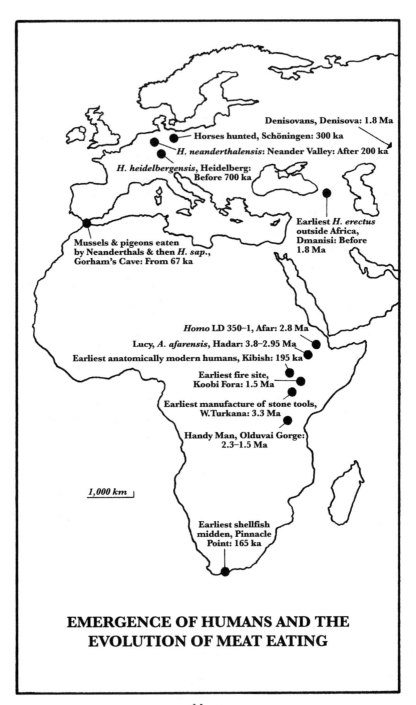

Denisovans, Denisova: 1.8 Ma

Horses hunted, Schöningen: 300 ka

*H. neanderthalensis*: Neander Valley: After 200 ka

*H. heidelbergensis*, Heidelberg: Before 700 ka

Earliest *H. erectus* outside Africa, Dmanisi: Before 1.8 Ma

Mussels & pigeons eaten by Neanderthals & then *H. sap.*, Gorham's Cave: From 67 ka

*Homo* LD 350–1, Afar: 2.8 Ma

Lucy, *A. afarensis*, Hadar: 3.8–2.95 Ma

Earliest anatomically modern humans, Kibish: 195 ka

Earliest fire site, Koobi Fora: 1.5 Ma

Earliest manufacture of stone tools, W.Turkana: 3.3 Ma

Handy Man, Olduvai Gorge: 2.3–1.5 Ma

1,000 km

Earliest shellfish midden, Pinnacle Point: 165 ka

# EMERGENCE OF HUMANS AND THE EVOLUTION OF MEAT EATING

Map 1

# 2

# *A Cooking Animal*

The idea that cooking makes us human is an old one. In 1785 the Scottish biographer and diarist James Boswell wrote: "My Definition of Man, is a 'Cooking Animal.' The beasts have memory, judgment, and all the faculties and passions of our mind, in a certain degree; but no beast is a cook. . . ." Boswell was writing before Darwin and so he was not making an evolutionary argument, but the idea that cooking is fundamental to our species is a conclusion that others have also felt in their very guts to be right. Gut instinct is generally frowned upon as a source of evidence in science, but guts are key witnesses in this matter, as we shall see.

Since no beast is a cook and, as Boswell said, we are cooking animals, the obvious question is, how and when did this habit evolve? Our great ape cousins are essentially vegetarians, living on leaves and fruits. Gorillas eat only plants, but chimpanzees will catch and eat animals when they can, though this is opportunistic behavior and they live mainly on fruit. Chimps can't cook, even though it has been argued that they are intelligent enough for the task. The common ancestor of chimps and ourselves must have been vegetarian, and so we meat-eating, cooking humans evolved by stages from vegetarian, indeed vegan, stock.

The yawning gulf between ourselves and other animals appears so large—not just in diet and cooking, but also in intelligence, language, brain size, and anatomy—because the intermediates along the evolutionary pathway that we unwittingly followed have been erased by extinction. We are the last surviving species of human in a world that once contained several others whom we ought to call sister and dozens more species who were ancestors and cousins. Collectively, we are the "hominins."

We are an African species. Charles Darwin deduced this, even before there was any fossil evidence, merely from the fact that those other great apes—chimpanzees and gorillas—are African. Nowadays, there is not only ample fossil evidence of our African origins, but support in the evolutionary history that can be read in our DNA. It is the mutations, or small changes in the genetic code, that enable us to reconstruct evolutionary trees by comparing DNA sequences. The process is very similar to the way in which the inheritance of surnames can be used to identify related individuals and map family trees.

Take my own name, Silvertown, for example. My paternal grandfather was born in Poland with the surname Silberstein. When he was four years old, the family migrated to England, where eventually my grandfather established a tailoring business. When the First World War broke out, German-sounding names were bad for business, so around 1914 my grandfather Anglicized his name to Silvertown. This mutation was an adaptation to local circumstances—something that happens all the time in evolution. Of course, genetic mutations are random, while my grandfather knew exactly what he was doing. I have a photograph of him standing proudly outside his shop under the signboard "Silvertown." His business thrived, the family grew, and today anyone with the surname Silvertown is (so far as we know) a descendant of my grandfather.

Other Silbersteins Anglicized the name too, but changed it to "Silverstone." In evolutionary terminology, each of the two mutations from Silberstein is known as a shared derived character. Shared derived characters can be used to reconstruct trees of descent—whether these are family trees or evolutionary ones. If your surname is Silvertown, this shared derived character tells us that you are a descendant of my grandparents Jack and Jenny. If your name is Silverstone, you belong to another branch of the family tree, and we have a more distant common ancestor. Mutation is an ongoing business. People frequently misspell my name as "Silverton." If I or one of my family decided to go with the flow and adopt this simpler spelling, that mutation would constitute a new shared derived character that would identify their descendants.

Now, returning to the wider family to which we all belong. When Darwin published his book *The Descent of Man* in 1871, the family album was just an empty book with a mirror for a cover. The first Neanderthal skulls had already been discovered, but their antiquity and significance were not then realized, so a hominin reunion at that time would have been a very lonely affair. Today thousands of hominin fossils have been discovered, and we even know the genome sequences of some of our more recent relatives. Since we are interested in what our ancestors ate and whether they cooked, what better way to find out than to invite them all to a fantasy dinner?

*El Día de los Muertos* is the Mexican festival when ancestors are celebrated and cemeteries become a picnic ground. Graves are decorated with flowers, and candy skulls and crossbones of sugar-frosted bread are traded as gifts. Our hominin reunion is going to be *Un Gran Día de los Muertos*—the biggest day of the dead, when representatives of our most ancient hominin ancestors will gather. Invitations have gone out, the word has gone the length and breadth of our African homeland, and from east to west across the planet, the news has spread that there is going to be a fiesta in the boneyard.

November first is here, and the day of the first grand hominin reunion dinner has arrived. Any hominin fossil with a tooth in her or his jaw is coming. Those known only from a scrap of bone and who cannot attend have e-mailed in their genome sequences. Now we need to provide a menu for the long-lost relatives. To make sure that every guest is catered for, we are going to have to ask each hominin that shows up: Who are you? When did you live? Where do you come from? And of course, what did you eat? Few of our guests would have been able to understand or answer these questions when they were alive, and now even the most intact skull can only reply with a rictus grin, but luckily many of the answers can be found by scrutinizing the arriving guests themselves. Though I would not advise that you try the same methods at home, as they include such intimate matters as cranial capacity, internal anatomy, and a microscopic inspection of the teeth.

First to arrive is our umpteenth-great-grandmother Lucy. Like all our oldest relatives, Lucy comes from East Africa. Her remarkably complete skeleton was found by Donald Johansen in the desert of Hadar in Ethiopia. She belongs to the species *Australopithecus afarensis* and was dubbed Lucy because the Beatles' song "Lucy in the Sky with Diamonds" was playing over and over in the camp when she was discovered. In life, Lucy was about the size of a chimpanzee, and she had a small, apelike brain that was only slightly larger than a chimp's; but the reason her discovery was celebrated was that she belonged to the first hominin species to walk upright like a human.

Though Lucy walked upright, we know from a remarkable piece of forensic deduction that she also climbed. Analysis of one of Lucy's arm bones shows that it was crushed in a fall from a great height. This fall was probably the cause of her death and suggests that although she could climb, she was not as adept at it as her more arboreal ancestors would have been. Her feet were, after all, made for walking.

The diet of Lucy and her kind, though mainly vegetarian, included a wider range of plants than is eaten by chimps, and it seems that *Australopithecus* species, of which there were several, were in general adapted to live in a wider range of environments than chimps do. *A. afarensis* had larger cheek teeth, smaller canines, and more powerful jaws than a chimp, suggesting that this ancestor did a great deal of chewing of tough food. The scientific consensus is that our own genus, *Homo*, arose from a species of *Australopithecus*, probably Lucy's own *A. afarensis*, which lived 3.8–2.95 million years ago.

Dear Lucy requires a booster seat to accommodate her small stature and, to be sure, she has the table manners of a chimp and no use for silverware, but how she will enjoy the crudités and the fruit salad! Perhaps Lucy will even steal some cooked food if she can get it from a neighbor, because experiments have found that great apes prefer this to raw food when they are offered a choice. In a remarkable study, the psychologist Penny Patterson raised a gorilla called Koko, whom she trained to communicate with her. She told primatologist Richard Wrangham what happened when she asked Koko what kind of food she preferred: "I asked Koko while the video was rolling if she liked her vegetables better cooked (specifying my left hand) or raw/fresh (indicating my right hand). She touched my left hand (cooked) in reply. Then I asked why she liked vegetables better cooked, one hand standing for 'tastes better,' the other 'easier to eat.' Koko indicated the 'tastes better' option."

Vegetarians leave little behind in the paleoarchaeological record to show what they ate—or rather, what they leave behind is very little. The characteristic shapes of very tiny grains of silica called phytoliths—which are part of the structure of leaves and when eaten can become lodged in teeth—can tell us something about what kind of plants Lucy ate. Meat-eating hominins, on the other hand, have obligingly left us not only the bones of what they ate, incised with the characteristic cut marks made by the stone tools that they used to butcher animals, but sometimes also the stone butchery implements themselves. The oldest bones with evidence of butchery actually come from Lucy's territory in Ethiopia. They are more than 3.39 million years old and show that the meat was stripped from them and that they were broken open to get at the marrow. It seems that *A. afarensis* cannot have been completely vegetarian and could process meat, not just gnaw at bones.

Until very recently, it was believed that the manufacture of stone tools was a strictly human skill and that all that the hominins before *Homo* could manage was to use some handy rocks to bash at bones and scrape at carcasses. However, in 2015 a paleoarchaeological site was discovered in West Turkana, Kenya, where stone tools were manufactured 3.3 million years ago (Ma), at least half a million years before the first species of *Homo* appeared. Elsewhere in East Africa, hominins living in Ethiopia 2.5 Ma were gutting, filleting, and perhaps also dismembering and skinning large animals. Collectively, these ancient remains of butchery take meat-eating way back before our own species, *Homo sapiens*, evolved just 200,000 years ago, to a time even before the first human (*Homo*) species evolved from *Australopithecus* around

2.8 Ma. So, humans are anciently meat-eating omnivores, and our earliest *Homo* ancestors went about butchering animals with gusto—indeed, as if their very lives depended on it. But who were they?

If we are arranging the place settings at the hominin family gathering in order of seniority, then the empty chair being kept at the table for the first species of *Homo* should be placed between Lucy, representing *Australopithecus afarensis*, on the one side and a recognizably human species called *Homo erectus* on the other. If the first human was intermediate between these flanking species, then by comparing them we can say he must have been bigger and brainier than *Australopithecus*, but what else was different and how many chairs must we reserve at the table to bridge the gap? Several species that are candidates to fill the space have been kept loitering in the lobby while paleoanthropologists have sought to place them correctly. One of the candidates is *Homo habilis*, or "Handy Man." The first fossil of this species was discovered and named in the 1960s from two pieces of skull and the bones of a hand that were uncovered lying next to some stone tools. Perhaps the scene of the first recorded kitchen fatality?

The original Handy Man fossil was only 1.8 million years old, but older fossils of this species have recently been identified, pushing the origin of *H. habilis* back to 2.3 Ma, much nearer the 2.8 Ma date when *Homo* is thought to have evolved from *Australopithecus*. This fossil indicates that *H. habilis* had a jaw more similar to *Australopithecus afarensis* and a cranial capacity more like *Homo erectus*, so it would look quite at home seated between the two. Judged by his teeth alone, Handy Man chewed with the same vigor as Lucy, but there is also a contender to squeeze into the space between the two of them.

In 2013 a new fossil jaw was discovered by Ethiopian anthropologist Chalachew Seyoum that looks like it belongs between *Australopithecus afarensis* and *Homo habilis*. It has been dated with remarkable accuracy to within just 5,000 years of 2.8 Ma, and while it had teeth with some human features, the shape of the jaw was like *Australopithecus*. This fossil has been given the unprepossessing name LD 350-1—better suited to an automobile license plate than a member of the family, you might think, but for the moment this species, which is neither *A. afarensis* nor *H. habilis*, has no other name. This new and very probably first member of our genus *Homo* was discovered just 30 kilometers from Hadar, where Lucy came from, and only 40 kilometers from the site of the oldest stone tools.

So, we have pinpointed within the compass of a couple of days' walk where in Africa hominins became human and began to butcher and eat meat. This is something to get slightly more excited about than the historic location of the first McDonald's hamburger restaurant. But, so far, all our guests at the hominin reunion are eating their food raw. Poor LD 350-1 wears a forlorn expression on his face as he toys listlessly with his name tag and chews for hours on a bloody steak. His dining companion, Handy Man, has cut up his meat with a stone blade that he has been fashioning for days.

As anticipated, *Homo erectus* has now arrived. As he enters, we can see that he stands only 1.3 meters (4 feet 3 inches) tall, though his body proportions are similar to our own. He has brought a stone hand ax to the party and looks like he could be trouble. Will he be offended if we offer him the same dish of raw meat that his companions are eating? Or will he just demolish the furniture and build a fire on which to cook it? A surreptitious look at his teeth might give us a clue. The earliest *Homo erectus* had large cheek teeth like those of their ancestors, *Homo habilis* and *A. afarensis*, but later *H. erectus* fossils reveal that over time they evolved smaller teeth that were suited to a softer diet requiring only half as much chewing. This suggests that *H. erectus* became an advanced practitioner of food preparation, maybe even a cook.

By 1.95 Ma, hominins, presumably *H. erectus*, living in the Turkana Basin in northern Kenya were butchering animals as challenging as hippo, rhino, and crocodile and also eating fish and turtles. However, we can be fairly certain that *Homo erectus* and his meat-eating forebears did not dine exclusively on meat. Any viable diet must provide energy as well as protein, and while lean meat supplies plenty of the latter, it is a poor source of calories because digesting protein and converting some of it into glucose consumes energy and liberates little. People who obtain more than about a third of their calories from lean meat quickly develop "rabbit starvation," a condition suffered by early American explorers who tried to survive only on the small animals they could catch. Eating just lean meat provides insufficient calories, causing people to eat yet more meat, if that is all the food available, in a vain attempt to satisfy their hunger. This then leads to meat toxicity.

Meat consumed to excess becomes toxic because the surplus amino acids produced when protein is digested overload the capacity of the liver to remove them. The liver turns excess amino acids into urea, which is then removed from the bloodstream by the kidneys, but the kidneys can also become overloaded by excess urea. These problems can be avoided if there is a sufficient proportion of fat in the diet because this supplies the missing calories, supplementing the demand for glucose and helping to satisfying hunger before too much meat is eaten. Adult Inuit are able to survive just by eating animals because the Arctic mammals they live on contain a great deal of fat, but children need some plant foods. However, the wild animals of the African savanna where *Homo* evolved mostly produce lean meat with little fat. Consequently, early *Homo*, having evolved from largely vegetarian ancestors could not cope with unlimited meat consumption in the way that true carnivores such as cats that are adapted to such a diet could.

Very probably early *Homo* obtained its main source of energy from the same sources as its ancestors—the carbohydrates of plants. Even today, most of the carbohydrate in our diet comes from plants, though from farmed sources such as wheat, corn, rice, yams, and potatoes. Surviving tribes of hunter-gatherers in Africa who lead lives that are probably similar to those of our long-ago ancestors obtain up to a third of their daily energy requirements from tubers, bulbs, seeds, nuts, fruit, and other wild plant sources. These are sources that would also have been available in Africa, 2 or 3 million years ago.

Direct fossil evidence of the plants that early hominins ate has not survived, but there is circumstantial evidence that they probably obtained carbohydrates from the underground storage organs of plants. For example, analysis of the tooth enamel of *Australopithecus bahrelghazali*—a species that lived on the shores of Lake

Chad in Central Africa when *A. afarensis* inhabited East Africa—has yielded chemical evidence that this hominin obtained as much as 85 percent of its calories by eating tropical grasses or sedges. Since the leaves of such plants are tough and unrewarding, it is most likely that *A. bahrelghazali* ate the fleshy stems and underground parts that are packed with starch. Indeed, both humans and baboons today eat the tubers of sedges such as the yellow nutsedge *Cyperus esculentus*. This species was widely cultivated in ancient Egypt for its tasty, nutritious tubers that are rich in both oil and starch and that may be eaten raw or cooked. Yellow nutsedge is grown as a crop in Spain, while elsewhere it is so prolific and persistent a plant that it ranks high among the world's worst weeds. A single tuber planted in an experiment in Minnesota produced more than 1,900 plants bearing almost 7,000 tubers in just twelve months!

Nutsedge tubers have a tough outer skin that would have presented a problem for a hominin lacking the right dental equipment. Could the stone flake tools so abundant at early hominin sites have been used to skin tubers? To find out if they could, 2 million-year-old quartz flake tools with the scratches and marks left on their sharp edges by their ancient use were compared with modern versions made from the same quartz at a site in southern Kenya. The modern, experimental flake tools were used to simulate the processing of different animal and plant foods, creating patterns of damage on their cutting edges that were characteristic of the different uses.

This experiment found that patterns of damage to the new tools made by skinning underground storage organs covered in grit, just as they came from the ground, matched some of the marks on the original tools, suggesting that those had also been used for that very purpose. If this were a classic detective mystery, we would conclude that hominins living 2 million years ago and more had the motive, the means, and the opportunity to include underground storage organs as a major component of their diet. The motive was the need for a source of carbohydrate, the means was the stone tool technology (if not the teeth), and the opportunity was the abundance of plants of the right type where they lived.

As we continue our inquiries in to how to feed our *Homo erectus* guest, we shall discover that he belongs to one of the most well-traveled hominin species. *H. erectus* migrated out of Africa, as our own species did, but more than 1.7 million years before us. The earliest human fossils found outside Africa belong to *H. erectus* and were discovered at Dmanisi in the Caucasus Mountains of western Asia. The Dmanisi fossils resemble early *H. erectus* from Africa and include the most complete early human skull known. These fossils date to around 1.8 Ma, showing that *H. erectus* entered Eurasia very soon after the species evolved in Africa. The range of the species expanded rapidly, extending from the Mediterranean in the west all the way to China in the east.

While we can be sure that *H. erectus* was an omnivore, eating plants as well as animals, their fossil remains are frequently found in association with those of elephants, on which they may have been particularly dependent. Elephants were hunted for food, and the fat as well as the meat from their huge carcasses would have been nutritionally important. Elephant bones and ivory were used to make tools, and wherever *H. erectus* roamed, there was always one species or another of these enormous herbivores available to supply a reliable source of prey. When the elephants disappeared from the eastern Mediterranean 400,000 years ago, so, too, did *Homo erectus*. In fact, almost wherever a human species appeared on the map in the last million years, the local elephant species went extinct.

So, if we have one of the later-model specimens with the smaller cheek teeth and the larger brain, it looks like we would be safe to offer our *Homos erectus* an elephant steak with a side of skinned nutsedge tubers for dinner. But will he send the order back to the kitchen to be cooked? There is good reason to think that he might, though direct evidence that *H. erectus* ate cooked food is surprisingly hard to come by. Fire sites, butchery, stone tools, and human fossils provide only circumstantial evidence of cooking. You may find the ashes of an ancient fire in a cave, but how can you be sure it was deliberately lit and not ignited by wildfire? The fire site may contain animal bones, but how do you know that the meat on them was cooked and eaten? If you are not of so skeptical a mind, then there are fire sites in Africa containing burnt animal bones, some even with butchery marks on them, that suggest the first cookouts happened as long ago as 1.5 Ma.

Luckily, because eating habits have so strongly shaped human evolution, we have biological as well as paleoarchaeological evidence that bears upon the case. Richard Wrangham, a primatologist at Harvard University, has assembled the evidence in his book *Catching Fire: How Cooking Made Us Human* into a convincing case that cooking was instrumental in the evolution of larger-brained *Homo erectus*, which he believes was the first human species to cook, 1.5 Ma. Wrangham points out that compared to chimpanzees, *Homo* species, including *H. erectus* and ourselves, have small mouths, weak jaws, small teeth, small stomachs, short colons, and altogether less gut. All these features of head and abdomen are adaptations suited to the energy-dense, soft food of a cooked diet.

We do not have direct evidence of the guts of *H. erectus*, of course, but from the size and shape of his rib cage, we know that he did not sport a belly big enough to accommodate the capacious guts of a raw-food herbivore. Australopithecine Lucy had the raw vegetarian diet that is the norm among primates, but humans are not equipped to process large amounts of bulky, fiber-rich, energy-poor foods. If our diet had not changed during evolution, a primate of our size would need a colon more than 40 percent larger than ours in order to provide the capacity required to digest raw plant food. People who try to live on this kind of diet and do not cook their food lose weight in an unsustainable fashion. Surviving for any length of time on a diet of only raw plant food, as other primates do, is impossible for us.

Now, surveying the guests so far assembled at the hominin fiesta, on the one side we can see the pre-cooking ancestors of the cooking animal, and on the other side we know where evolution led—but what is less clear is exactly when the big dietary change caused by cooking happened and why. It seems highly likely on anatomical grounds that the first cook was *Homo erectus*, but how early in the long history of this ancestor did cookery begin?

There is genetic evidence that a gene called *MHY16* that strengthens the jaw muscles of non-human primates was lost from the human lineage more than 2 Ma. Perhaps the earliest *Homo erectus* were already cooking at that time, and powerful jaw muscles were becoming redundant, or even a breakage hazard for teeth that were getting smaller. The answer to exactly when cooking began is likely to become clearer as more fossil and paleo-archaeological evidence is discovered. By comparison with the mystery of *when* cooking began, the question of *why* has a much clearer answer. Cooking increases the digestibility of food, enabling us to extract more energy from it, and it inactivates many toxins, thereby opening up new vistas of possibility in hominin evolution.

The tuber of a potato or a nutsedge is a well-armed vault in which the plant that makes it stashes an energy supply for future use in its own growth and reproduction. As you might expect, these stores of precious energy are protected against outside attack by a battery of defenses. First, they are buried out of sight so they have to be discovered and dug up. Then, like nutsedge, they may have a tough skin, or like manioc (cassava) they may be dosed with toxins that render them inedible without processing. The starch in tubers is also packed hard, so that it is inaccessible to digestive enzymes in the gut. In children in particular, potato that has been inadequately cooked can pass right through the gut in intact pieces. Finally, starch molecules are locked in crystalline blocks inside tiny granules that are so small that they cannot be broken open by grinding between teeth or even between stones. Cooking undoes most of a tuber's defenses, destroying toxins and enzyme inhibitors, softening the tissue and busting open starch granules as the starch changes from a dry crystalline to a moist gelatinous form that is accessible to the digestive enzymes able to break it down. Meat and fat also yield nutrients, energy, and flavor to the cooked food in quantities that, when raw, can only be got in a lion's stomach.

Wrangham argues that cooking made us human because it gave us the energy needed to power a large brain. The single most important trend in human evolution is the steady increase in brain size that has taken place over the last 2 million years. Our brains are now three times bigger than that of any other primate, though absolute size is not everything. Cows have big brains too but are not that bright. Large, smart brains unlocked the possibility of uniquely human capabilities such as complex language, abstract thinking, and all that flowed from that. Brains are very energy-hungry organs. The human brain is only about 2 percent of body weight, but uses fully 20 percent of the energy consumed in a resting state. Most of this energy is used at the electrical junctions called synapses, which connect nerve cells to one another and are the cornerstones of brain function.

Weight for weight, guts are about as energy-hungry as the brain, but while our brains are much bigger than the norm for a primate of our size, our guts are much smaller. By economizing on guts, evolution spared energy for splurging on bigger brains. Wrangham's hypothesis is that by increasing the energy value of our food, cooking made it possible for smaller guts to supply the burgeoning requirements of brain evolution. If you think of the gut as a fuel tank, then cooking increases the octane of the fuel, but alongside this, humans benefit from a faster-running engine too. The latest research comparing metabolic rates in great apes and humans has unexpectedly found that our metabolic rate is 27 percent higher than that of a chimp. So, we not only have higher-octane fuel, but we burn it faster too. Pound for pound, the human energy budget is bigger than a chimp's. What do we spend the extra on? Just think!

Perhaps the most persuasive evidence that we truly are the cooking animal is that brain growth and cooking do seem inextricably linked. During the evolution of humans, our guts shrank at around the same time that our brains grew. *Homo erectus* is the embodiment of this trend, and if Wrangham is right, then by now our guest will be pounding on the table and bellowing for his cooked dinner. Never have the dead been heard to make such a racket.

With the quandary of what to feed *Homo erectus* now resolved and our hungry ancestor quieted with mouthfuls of cooked grub, we can turn our attention to the next guest. A tall, strongly built hominin strides confidently into the room carrying a slender wooden spear, more than 6 feet (2 meters) long, tipped with a well-made stone arrowhead. This is *Homo heidelbergensis*, descended from the African branch of *Homo erectus*, but having a more modern appearance and a 30 percent larger brain than that species. The forehead is higher and the face is flatter, but it is still adorned with pronounced brow ridges and there is no chin. *H. heidelbergensis* emerged more than 700,000 years ago and so was the product of a lineage that had already undergone a million years or more of brain enlargement. As its scientific name suggests, the first *H. heidelbergensis* fossil was discovered near the German city of Heidelberg, but others were later found in Greece, Ethiopia, and Zambia. There are also putative fossils of *H. heidelbergensis* from India and China.

*H. heidelbergensis* is the first of our ancestors whom we can be reasonably sure could obtain fire whenever they needed it. The spear that our guest is carrying is made of spruce wood and is one of several found buried in mud at Schöningen in Germany. The Schöningen spears date from about 300,000 years ago, when the area stood on the shores of a lake that abounded with animal life. Elephants were present, though rare, and hominins mainly hunted and butchered horses, whose dismembered remains litter the site. A single horse kill would have fed a band of twenty or thirty people for two weeks, and their meals could have been garnished with the produce of local wild plants such as hazelnuts, acorns, and raspberries. For this particular relative, a medium-rare horse steak served with roasted acorns, followed by ripe hazelnuts crushed in a raspberry coulis sweetened with wild honey sounds about right, don't you think?

With *Homo heidelbergensis* happily settled at the table and that intimidating spear safely parked out of the way, we can turn our attention to this hominin's descendants and our final dining companions. Breaking with family tradition, two of these descendants did not evolve in Africa, but were the spawn of emigrant *H. heidelbergensis*. The most well-known of the two, who could have made it into the family album nearly 200 years ago if their antiquity had been realized when their fossils were first discovered in the nineteenth century, are the Neanderthals, *Homo neanderthalensis*.

The other is an extinct cousin we did not even know we had until 2010, when DNA analysis of a finger bone discovered in a Siberian cave produced a sequence that fits neither Neanderthals nor our own species. This DNA sequence, which proved to belong to a young girl, was sufficiently different that anthropologists have dubbed these hominins as a distinct species called the Denisovans, after the locality Denisova, where the finger bone was discovered. We have so little physical evidence of the Denisovans that they easily take the prize for the most ghostly presence at our reunion, even though it is a fiesta of the dead. Genome sequencing has revealed the shadowy presence of Denisovan genes in some present-day populations of our own species. These are telltales of an encounter that must have taken place more than 50,000 years ago between Denisovans and our species as we were en route to the peopling of Melanesia and Australia, where people today have inherited a little fraction of their DNA from this meeting.

We will leave an empty place at the table for the Denisovans, marked with a collection of ornaments made from the teeth of fox, bison, and deer found in the Denisova cave and that might just have belonged to the little girl who died there. It will probably not be long before more Denisovan fossils turn up. Meanwhile, I hear the heavy tramp of large hominin feet upon the stair. Make way for our final guests, *Homo neanderthalensis*!

There enter a man and a woman with a babe in arms. These are modern-looking people who, if they paid a visit to the hairdressers and a clothes store, you might then pass in the street with only a sidelong glance at their strongly muscled appearance, unusually large noses, and chinless features. The Neanderthals were natives of the Northern Hemisphere, not Africans like us, and were adapted to a cool climate with dark winters. The genome of one of the first Neanderthal to be sequenced revealed that he was a redhead. Though we both descended from *Homo heidelbergensis*, the Neanderthals evolved from that species in Eurasia and we from its African population. A comparison of the genomes of our two species shows that we shared a common ancestor more than 500,000 years ago. The Neanderthals hung on in Europe till as recently 40,000 years ago, but their extinction was not without trace. There are Neanderthal genes in all human populations outside Africa. We also know a good deal about what the Neanderthals ate.

There are three principle sources of information about the Neanderthal diet: the calculus from their teeth that contains remnants of what went in, their fossilized feces that tell us what came out, and the bones and debris left, metaphorically speaking, on the side of the plate. Neanderthal cave dwellings are so littered with animal remains that the obvious conclusion would be that they subsisted mainly by hunting large animals and eating meat. However, unless it contained plenty of fat, such a high-protein diet would not have satisfied their energy requirements, especially since these were probably greater than ours because they were more muscular than us and had slightly larger brains than us too. Chemical analysis of 50,000-year-old Neanderthal feces supports the notion that they ate a lot of meat, but also shows that they ate greens as well. Other evidence supports this conclusion too.

The formation of dental calculus on the teeth is like a living fossilization process that samples the contents of the mouth, potentially over a whole lifetime. It starts with the deposition of bacterial plaque on the teeth. Over time, this becomes mineralized by the deposition of calcium phosphate, which is present in saliva at a supersaturated concentration. The biological function of calcium phosphate in saliva is to repair dental enamel, but a side effect is that plaque becomes mineralized, trapping food particles inside a crystalline matrix that preserves them for millennia.

Dental calculus taken from Neanderthal teeth has been found to contain phytoliths from a wide range of plants including dates, underground storage organs, and grass seeds, plus cooked starch grains and even smoke particles. Short of finding a Stone Age recipe book, this is the clearest evidence available that Neanderthals actually cooked and ate plants. Plant remains decay very easily, but if they happen to get charred in a fire, this can preserve them and provide another source of evidence. Such burnt remains from a cave on Mount Carmel in Israel indicate that Neanderthals there gathered almonds, pistachios, acorns, wild lentils, and the seeds of wild grasses and many plants in the pea family. Neither chicken soup nor falafel had been invented yet.

The latest evidence suggests that the breadth of the Neanderthal diet was not very different from that of our own species at a similar time. Neanderthals did not just subsist on large animals, though these were important, but also cooked and ate shellfish and occasionally smaller game such as rabbits, tortoises, and birds. Gorham's Cave in Gibraltar, a rocky promontory in the south of the Iberian peninsula that overlooks the entrance to the Mediterranean, is one of the last sites that Neanderthals occupied—perhaps even their last redoubt. Rock doves (pigeons) nest on the cliffs around the cave to this day and were taken for food and cooked by the Neanderthal inhabitants as a regular harvest from 67,000 years ago until they, but not the pigeons, disappeared. Our own species then occupied the cave and continued to eat the local pigeons for thousands of years more.

Each of our guests at the hominin reunion has now been served their customary diet, insofar as we can tell what that was, every skull wears a satisfied grin and there is ghostly belching in the dining hall. Five million years ago our ancestors were probably largely vegetarian, by 3.3 Ma they were fashioning stone tools and eating meat, and by 1 Ma, probably earlier, they were cooking their food. From this history we learn that evolutionary change is gradual and that the origins of current habits such as toolmaking and cooking that we usually regard as novel and peculiar to us can actually have deep roots in the hominin lineage. That lineage is ancient, though our own species is very recent.

Now our own parvenu species is ready to join the feast. The continent of Africa, the source of *Homo sapiens*, lies just nine short miles across the Strait of Gibraltar from Gorham's Cave, where the Neanderthals ate their last supper of pigeon, but when we departed Africa we neither crossed the strait nor dined on pigeons en route. We dispersed from Africa by an altogether more circuitous route and dined on something quite different.

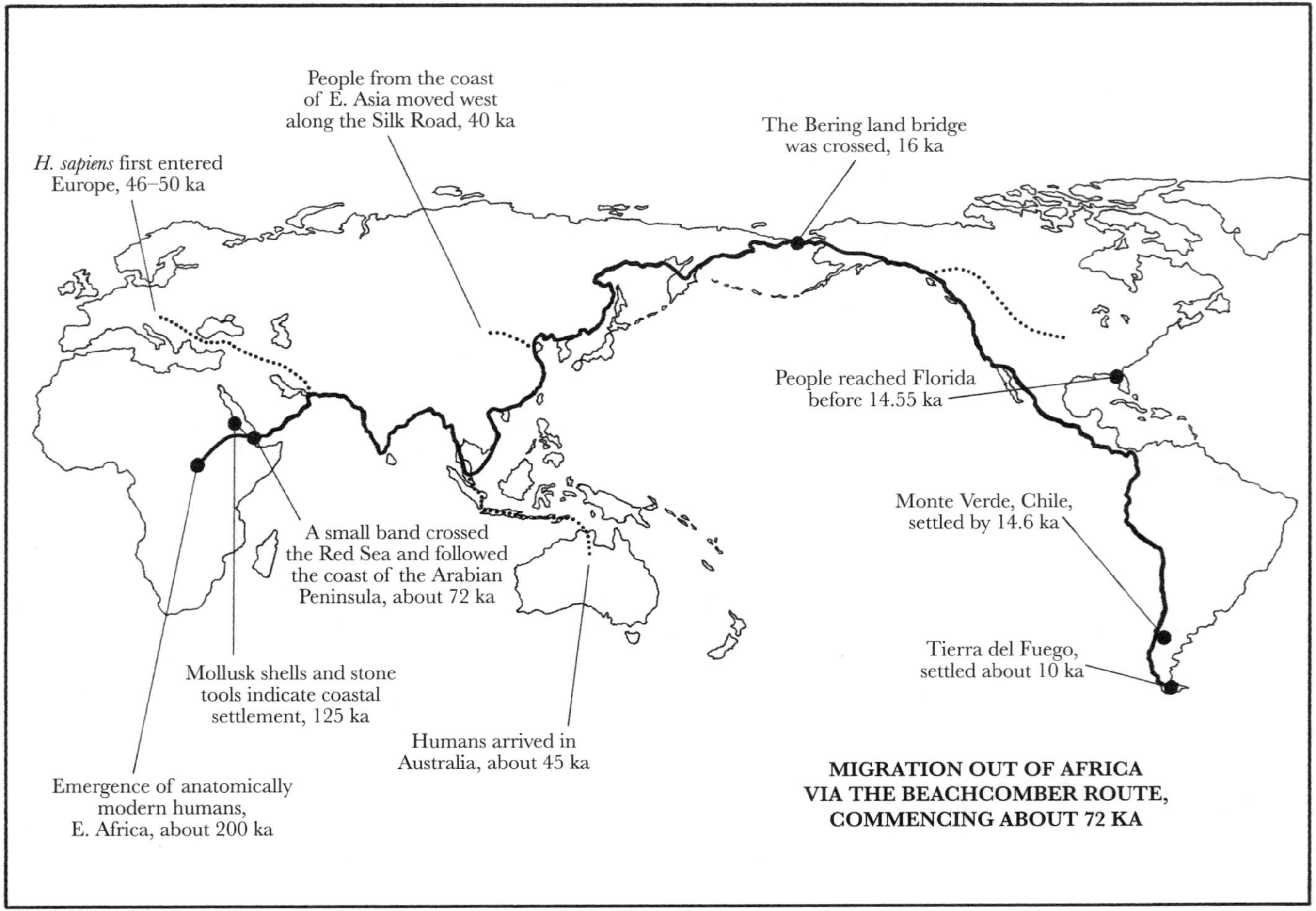

The Bering land bridge
was crossed, 16 ka

People reached Florida
before 14.55 ka

Monte Verde, Chile,
settled by 14.6 ka

Tierra del Fuego,
settled about 10 ka

People from the coast
of E. Asia moved west
along the Silk Road, 40 ka

Humans arrived in
Australia, about 45 ka

A small band crossed
the Red Sea and followed
the coast of the Arabian
Peninsula, about 72 ka

H. sapiens first entered
Europe, 46–50 ka

Mollusk shells and stone
tools indicate coastal
settlement, 125 ka

Emergence of anatomically
modern humans,
E. Africa, about 200 ka

**MIGRATION OUT OF AFRICA
VIA THE BEACHCOMBER ROUTE,
COMMENCING ABOUT 72 KA**

Map 2

# 3

## *Shellfish—Beachcombing*

In 1440 the anonymous author of a medieval "Boke of Kokery" recorded cooking instructions for a dish of mussels. The recipe was written in Middle English, and although the spelling is unfamiliar to us, the sound and meaning of the words penetrates the six centuries that have passed. "Take and pike faire musculis," says the monk. "And cast hem in a potte." Add "myced oynons, And a good quantite of pepr and wyne, And a lite vynegre." You will know when they are ready "assone as thei bigynnet to gape, take hem from ye fire, and serve hit forthe with the same brot in a diss al hote." The essential ingredients of this dish have not changed: clean mussels, minced onions, pepper, wine, a little vinegar.

Mussels are a food almost as timeless as mother's milk. People have been eating mussels, raw or cooked, for at least 165,000 years, probably much longer. Our siblings the Neanderthals also ate mussels and therefore quite probably our common ancestor more than half a million years ago did too. Hominins could have been eating shellfish for a million years, perhaps more. This is a modest claim since many monkeys and apes have been observed to take fish and shellfish when they are obtainable.

Mounds of discarded seashells punctuate the historical routes followed by our species as we traveled along the coastlines of our planet. From the Arctic in the north to the southern coast of Africa and the tip of South America, shells cast aside by people who gathered mollusks between the tides are testament to the diet of generations. Seafood is rich in omega-3 fatty acids that are crucial to brain development, so this source of food may have been nutritionally essential in our evolution. Essential nutrients are the vitally important compounds, like certain amino acids, that our cells cannot manufacture for themselves and that therefore must be acquired through food.

The oldest seashell midden is from the Middle Stone Age in Africa, 165,000 years ago, where some of the earliest modern humans inhabited a cave with a distant view of the Indian Ocean. From the trash left behind, we know that they were hunter-gatherers who dined on species of shellfish that can still be found in the area today. These included several species of mussel, lots of limpets, and a large sea snail with a shell that has thick coils like a turban and is called in Afrikaans "Alikreukel," a half dozen of which make a very fine supper.

This remarkable cave is at Pinnacle Point in the Cape of South Africa and was discovered by Curtis Marean, an anthropologist from Arizona State University. He wrote that he did not stumble on the cave by mere chance, but was guided to the area by the knowledge that within a few tens of thousands of years after the birth of our species in Africa some 195,000 years ago, a glacial cooling and drying of the climate made most of Africa too inhospitable for human survival.

The genetic impact of the population collapse triggered by this glacial period is still engraved in modern genomes and suggests that the human population was reduced from 10,000 breeding individuals to perhaps only hundreds. Marean deduced that these survivors, from whom we are all descended, might have found refuge in the Cape of South Africa because the influence of the surrounding oceans moderates the climate there. The Cape could have provided two food sources that would have been unaffected by the cold dry conditions that reduced the game that hunter-gatherers depended upon for food elsewhere: seafood and the abundant bulbs of plants that grow in the Cape's unique flora. Those bulbs would have been the Middle Stone Age equivalent of the onions that are used in cooking a dish of mussels today.

Modern sea level is much higher than it was 165,000 years ago, when ice locked much water on land, and so any inhabited caves that were near the coast then would since have become submerged or scoured free of any archaeological deposits by the waves. Evidence of human habitation in the Pinnacle Point cave survived because it was farther inland and high up in a cliff. The archaeological deposits found there suggest that the cave was occupied intermittently, perhaps only in periods when fluctuations in sea level brought the coastline near enough to provide convenient access to shellfish.

Marean suggested that most of the archaeological evidence for human habitation in the area must be buried in sediments offshore. Such evidence from a later date has been found farther north, in the shallow waters of the Red Sea, off the coast of Eritrea. There, excavation of a coral reef has uncovered hundreds of stone tools embedded in coral that grew over them as sea level rose. These finds have been dated to 125,000 years ago and occur among the remains of thirty-one species of edible mollusks, including oysters and massive beds of mussels. Two species of edible crab were also present with the stone tools that would have been used get to their delicious meat.

The Red Sea coast was the waiting room for the human departure from Africa. How many unsuccessful forays were made from there we do not know. There may have been many. We do know that the descendants of one of these forays made it all the way to China, where teeth of anatomically modern humans dating to 100,000 years ago have been discovered. However, these pioneers appear to have died out because the genetics of modern people outside Africa show that everyone descends from a later excursion.

Our species (*Homo sapiens*) was confined to the continent of Africa for 50,000–60,000 years after the first inhabitants of Pinnacle Point dined on shellfish, but around the same time that Pinnacle Pointers were picnicking, Neanderthals (*Homo neanderthalensis*), who appear to have been better adapted to cold climates than we were, had already spread across Europe and were living around the southern coast of Spain. Many of the mussel shells retrieved from Neanderthal caves there are burned on their outer surface, indicating that they were roasted in a fire.

Was the art of seafood cookery more advanced among Neanderthals than in our own species? If there was a Stone Age cook-off between our two species, it would have had to wait till around 100,000 years ago, when *H. sapiens* from North Africa traveled eastward along the southern shore of the Mediterranean and appeared in the what is now Israel. Southwest Asia, which very much later was to become the birthplace of agriculture, was already home to Neanderthals, and whether we were bested by them in competition for food or merely slain by their cookery, *H. sapiens* did not survive.

Our first successful exit from Africa took place about 30,000 years later (72,000 years ago). This was also a coastal migration provendered by seafood, but this time it took a southerly route across the mouth of the Red Sea and then around the coast of the Arabian peninsula into India (map 2). Why did people so plentifully provided with food in the Red Sea make the trek out of Africa? Although we do not actually know the answer to this question, it might have been the pressure of a growing population upon the coastal food resources. What we do know is that this migration was the singular beginning of the peopling of the globe by our species.

The event was "singular" in both senses of the word: it was remarkable and it was unique. The entire human population outside Africa—some 6 billion of us—appears to be descended from the small band of people who, one fine day about 72,000 years ago, crossed the Red Sea from the Horn of Africa into the Arabian peninsula. We know this because it is recorded thus in our genes. The population of Africa is richly diverse in its genetics, with many differences among its peoples. In comparison, the rest of the world is genetically uniform, all sharing the limited sample of Africa's human genetic diversity that was carried by those, perhaps few hundred individuals, who made that remarkable exodus. The farther our journey took us from Africa, the more of our native genetic diversity we lost. This indicates that each step of the migration was taken by small groups of individuals who broke away, traveled, set up camp, and established their own settlements, which then in turn spawned a breakaway when the number of settled people became large.

The journey that began with the exodus from Africa around 72,000 years ago continued mainly along coastal routes, provisioned by the seafood that had served us so well along the coasts of our home continent. Traveling around the coast of India, we reached the Australian continent some 45,000 years ago, with that date of arrival marked by the appearance of the familiar middens of shellfish.

The genetic evidence tells us that along the route of the coastal path, bands of people broke away at intervals and struck inland. The descendants of one of these bands first entered Europe 45,000–50,000 years ago. The Asian hinterland was colonized by later breakaways when 40,000 years ago, people traveled from the coast of East Asia, inland and back toward the West, becoming the first travelers along the path that would become the famous Silk Road between China and Europe.

By about 16,000 years ago, the coastal migration around the Pacific Rim had traveled north as far as Siberia. While the continent was still covered in ice, the coast had become ice-free, providing a route into northwest North America. All Native Americans from Alaska to Chile are descended from these first colonists from Asia. From this northern point of arrival, people spread out and appear to have taken multiple paths into North America. We know that Florida was populated by 14,550 years ago because butchered mastodon bones of this age have been discovered there. Others followed the Pacific coast, reaching Chile in South America more than 14,600 years ago. To this day, with its 4,000 miles of coastline, Chile is the capital of shellfish gastronomy, offering abalone almost large enough to be carved like a joint of beef.

Ultimately, the Pacific coast migration reached the southern tip of the South American continent at Tierra del Fuego, possibly around 10,000 years ago. All physical signs of the first Fuegians have been obliterated by the sea and by a volcanic eruption that occurred in 7750 BP, but we have a vivid firsthand account from a later date of what life must have been like at the very limit of the human journey from Africa. During his voyage on the *Beagle*, Charles Darwin visited Tierra del Fuego, and on Christmas Day 1832 he noted in his diary:

> The inhabitants, living chiefly upon shellfish, are obliged constantly to change their place of residence; but they return at intervals to the same spots, as is evident from the piles of old shells, which must often amount to many tons in weight. These heaps can be distinguished at a long distance by the bright green colour of certain plants, which invariably grow on them.

Darwin pitied the Fuegians, dressed in scraps of sealskin or entirely naked, exposed to wind and rain and sleeping upon the wet ground in near freezing temperatures. "Whenever it is low water, winter or summer, night or day, they must rise to pick shellfish from the rocks."

Recent exploration of the Beagle Channel, named after Darwin's ship, has found that shell middens are to be found everywhere, even on the smallest places that are accessible to a canoe. The majority of shells belong to various species of mussel, and the biggest middens are up to 3 meters deep and 50 meters across, representing intense exploitation of mussel beds over very long periods of time. Archaeological excavations have found that people have been living on shellfish in the area for over 6,000 years.

For us—who possess the privilege of choice about what we eat, how much, and when—gathering shellfish on the shore is a seaside amusement, but the Fuegians often starved when stormy weather made it impossible to gather shellfish or to hunt seals from their canoes. In human history, this must often have been the plight of gatherers upon the shore. Shellfish were famine food and only very recently have they become an expensive delicacy. They sustained us through the hard times in Africa and for 60,000 years, until the invention of agriculture, they fueled our coastal travels around the globe. With agriculture came the new technology of plant and animal domestication and a revolution in diet with consequences every bit as momentous as the shift from a vegetarian to an omnivorous diet or the invention of cooking.

# 4

## *Bread—Domestication*

When it was first made, bread represented something new in the history of eating: a processed food. A field of wild grasses does not obviously promise a meal for the taking in the way that mussels, bulbs, fruit, or wild animals do. The seeds of cereal grasses must be gathered, threshed, and winnowed to separate the grain from the chaff, ground to flour, mixed with water to make dough, fermented, and then cooked before they are eaten. But, such is the reward in flavor and nutrition for all this effort, that bread became synonymous with food itself.

By Roman times, wheat and barley had been staples of the diet in Europe and southwest Asia for thousands of years. The cities of ancient Rome and Greece and the pyramids of Egypt were built on bread just as surely as they were built of stone. Thanks to archaeology, we know just what that bread was like and even how it was made.

Drying preserves food and other organic materials remarkably well because the microbes that cause rot and decay cannot live without moisture. The arid climate of the Egyptian desert has preserved hundreds of loaves of bread that were placed in ancient tombs 3,000–4,000 years ago for the postmortem feasting of royalty. These loaves were made mainly of a domesticated wheat species called emmer, with the occasional addition of fruit. Emmer is no longer grown as a crop, but it is an ancestor of both the main types of wheat that are grown today: durum wheat, which is particularly suited to making pasta, and bread wheat, which evolved from a cross between domesticated emmer and a wild species of goatgrass.

Archaeological excavations have uncovered the villages where the laborers who built the pyramids lived, and these reveal that workers as well as royalty ate wheat bread. Each small dwelling in the crowded village was individually equipped to mill emmer and to home bake loaves. The economy of ancient Egypt operated through barter, with value typically calculated in quantities of grain and the amounts of bread and beer that could be produced from it. Laborers were paid near-subsistence quantities of grain, but high officials received huge quantities that were far greater than anyone could eat.

Whether living or dead, a king does not do his own baking, so to supply him in the afterlife when the bread supply ran out, the priests who furnished the tomb of King Nebhepetre Mentuhotep II (d. 2004 BCE) gave him a miniature model of an industrial-scale bakery. The model, which now resides in the British Museum in London, has thirteen small figures kneeling before saddle-shaped stone querns in which they are milling wheat with a stone held between two hands. The use of coarse granite grindstones introduced rock particles into the flour, making a gritty bread whose effects can be seen in the severe tooth wear of Egyptian mummies. An eternity eating ancient Egyptian bread would require an inexhaustible supply of dentures.

In front of the baker's dozen of millers is a rank of figures kneading the dough, and behind them are three cylindrical ovens, each tended by a baker. Thanks to a frieze in another Middle Kingdom tomb, we can eavesdrop on a conversation that might have taken place in such a bakery. The frieze is in the tomb of Senet, a female relation of the most senior court official serving the king that succeeded the one buried with the model bakery. Women only rarely received tomb burial, but the interior of Senet's tomb at Luxor is elaborately decorated with scenes of life along the Nile, including fishing, hunting with dogs, the butchery and preparation of meat, and the production of bread and beer. The bakers chat to each other in hieroglyphs that, deciphered, read like a comic book dialogue. A woman grinding grain at a quern piously proclaims, "May all the gods of this country give health to my powerful master!" And a companion, her words censored by the decay of time, says something that ends ". . . this is for food." Perhaps she wants to make clear that her batch of flour is not to be used for making beer, which was made from both wheat and barley in those times. Men working in the bakery complain ". . . I am hard at work"; "None of you will give me a moment"; and "This firewood is green . . ."—with a hand raised to protect his face from the smoke and heat.

These scenes from ancient Egypt are so timeless that it may seem as though bread and agriculture have always sustained us, but they have not. Agriculture began between 12,000 and 10,000 years ago in southwest Asia. The very earliest evidence of agriculture in this region has been found in Anatolia in southeastern Turkey, and not long afterward it appeared throughout the area of southwest Asia called the Fertile Crescent (map 3). From Anatolia, this region arcs southward through present-day Lebanon, Israel, and Jordan to the Nile valley in Egypt, and from Anatolia eastward through northern Syria into Iraq, turning south through the land of ancient Mesopotamia, watered by the rivers Tigris and Euphrates. Clay tablets dating from around 4,000 years ago record that in Mesopotamia at that time there were around 200 kinds of bread, varying in the kind of flour used, what other ingredients were added, how the dough was made, how the loaf was cooked, and how the finished article was presented.

Emmer wheat appears to have been the very first crop to be domesticated, but it was rapidly joined by other species that made a group of eight or nine founder crops. In addition to emmer wheat, the first farmers in the Fertile Crescent also domesticated einkorn wheat, barley, lentil, pea, garbanzos (chickpeas), bitter vetch, flax, and probably fava bean (broad bean). To this day, ancestors of all the founder crops except one can be found growing wild in the Fertile Crescent. The exception is the fava bean, whose wild ancestor has not yet been discovered despite much botanical detective work. The wild fava bean might even be extinct.

It may seem a strange coincidence that so many wild species suited to cultivation were to be found in one place, but there is

## THE FERTILE CRESCENT IN SW ASIA

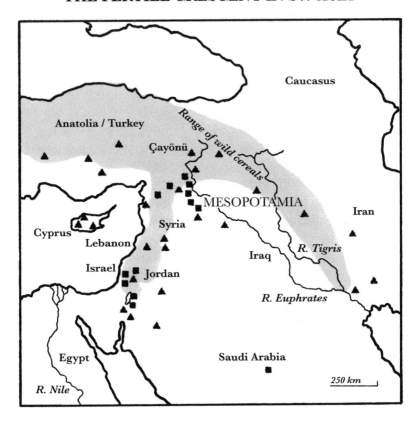

■ Archaeological sites with evidence of pre-domestication cultivation

▲ Neolithic sites with remains of domesticated crops

Map 3

a good evolutionary reason for this. The explanation is in the climate. Rainfall in the Fertile Crescent is highly seasonal and uncertain. A dry climate with uncertain rainfall favors the evolution of three characteristics in wild plants that make them particularly well-suited as the raw material for domesticated crops. The first characteristic is short life. Short-lived annual plants grow and mature quickly, setting seed in abundance before the plants die in the arid heat of the summer.

Plants with an annual life history are not just convenient to grow and harvest, but they are also generous with their bounty.

Fecundity is their second useful characteristic. Because annuals have only one chance to reproduce, they put a greater proportion of their available energy into making seeds than do perennial plants that can spread their reproduction over several years. Thus, all our grain crops are annuals, including those domesticated elsewhere, like corn and sunflower in the Americas, sorghum and pearl millet in Africa, and rice in Asia. Annual plants give the biggest bang for the buck, and through artificial selection this potential can be tuned to the maximum.

The third characteristic that made wild annual plants of the Fertile Crescent such good raw material for domestication was the relatively large size of their seeds. A dry climate favors the evolution of large seeds because for a seedling to survive when a seed germinates, it must produce a root that will supply the developing plant with water. In dry environments, roots need to go deep to find water, and a seedling can only make long roots if it has substantial food stores, which means a big seed.

The effulgent hue of a golden wheat field just before harvest is a sight that can now be seen in Australia, North America, northern Europe, southern Africa, India, or the Ukraine, but all are images of the original wild cereal fields of the Fertile Crescent. Fields of wild cereals containing emmer wheat, barley, and oat can still be found on rocky ground in Turkey, Israel, and Jordan. The American dean of crop evolution, Jack Harlan (1917–1998), visited Anatolia in southeastern Turkey in the 1960s, where he found huge natural stands of wild einkorn wheat growing on the slopes of Karacadag Mountain. As an experiment, he used a replica of an ancient sickle with a flint blade to see how much wild grain he could collect in an hour. Although some grain was lost as the wild seed heads broke up, Harlan managed to collect nearly 2.5 kilos (5 pounds) of wheat per hour. After threshing, this yielded only half the weight in grain, but the protein content of the wild wheat was 23 percent by weight, or 50 percent more than in some modern domesticated varieties.

Harlan calculated that a family harvesting wild einkorn could collect enough grain in just three weeks to last them a whole year if they followed the ripening of the wheat as it proceeded from the earlier-ripening lower slopes to the later harvest higher up the mountain. So abundant was this crop that Harlan asked: "Why should anyone cultivate a cereal where natural stands are as dense as a cultivated field? If wild cereal grasses can be harvested in unlimited quantities, why should anyone bother to till the soil and plant the seed?" The answer is that for a long time, gathering from the wild was probably indeed sufficient. This would explain why the archaeological record shows that cereal domestication took thousands of years. Eventually, however, as human population numbers increased, domestication and farming became a necessity.

Archaeological sites have yielded abundant evidence of how people began gathering the seeds of wild plants for food and then started to cultivate and alter them through a process of artificial selection. Wild emmer and wild barley were gathered 23,000 years ago by people living on the shores of the Sea of Galilee in Israel. Wild grasses including wheat, barley, and oats have seed heads that fall to pieces, or "shatter," when they are ripe, dispersing the seeds that they carry. Natural selection has endowed the young of all species with a means of dispersal because this improves their prospects of survival and reproduction. However, the mode of dispersal changes when plants are taken into cultivation and become domesticated. In this situation, the plants that multiply most are those producing seeds that are collected and then re-sown. Continual harvesting and re-sowing therefore select for plants with seed heads that do not shatter during reaping.

In the early stages of domestication, grains gathered from the wild are indistinguishable from cultivated ones in the archaeological record. As cereals became domesticated, genes that prevent seed heads from shattering began to increase in frequency through artificial selection. Non-shattering seed heads hold on to their seeds. These have to be broken away by mechanical force during threshing. When non-shattering seed heads were threshed in the farmyard, instead of breaking cleanly in the natural fashion of wild plants, seed heads fractured, leaving jagged edges. Therefore, a high percentage of jagged breaks, visible under a simple lens, is the signature of cereal domestication. Emmer wheat is the first cereal crop to appear in the archaeological record with this telltale sign of domestication.

The earliest site where such remains have been discovered in the Fertile Crescent is at Çayönü in Anatolia, in southeast Turkey. Emmer wheat was cultivated there about 10,000 years ago, but its grains were relatively small and strongly resembled those of wild plants. The village of Çayönü also cultivated peas, lentils, and flax. Though, at the time of writing, Çayönü is the site of the earliest positive evidence of cereal cultivation judged by the signature of domestication, other archaeological evidence suggests that agriculture must have been widespread in the area. Crop domestication probably began hundreds, perhaps thousands of years before its signature became unmistakable, during which time the various founder crops would have been evolving under domestication, interbreeding with their respective wild progenitors and being exchanged among farmers across hundreds of kilometers in the Fertile Crescent.

Non-shattering seed heads are the first sure archaeological sign that a cereal crop is domesticated, but domestication also selects for other characteristics in the crop that are different from the wild ancestor, particularly larger grain size and loss of seed dormancy. Once farming began to spread, the domesticated crops that were carried to new regions had to adapt to new climates. In his book on domesticated animals and plants, Charles Darwin commented that when the first European settlers arrived in Canada, they

> found their winters too severe for winter-wheat brought from France, and their summers often too short for summer-wheat; and until they procured summer-wheat from the northern parts of Europe, which succeeded well, they thought that their country was useless for corn crops.

Nowadays, so successful is wheat farming with cultivars that are adapted to the climate of Canada that in 2013 there were insufficient rail cars available to ship the bumper crop of 37 million metric tons. Such is the bounty of adaptation.

There are now hundreds of thousands of wheat varieties, most of them bread wheats, but all this diversity is built upon the foundation of two momentous evolutionary events. The first occurred between 500,000 and 800,000 years ago, when wild emmer evolved through hybridization between a species of goatgrass and a wild wheat. The second event was much more recent and produced the ancestor of all bread wheats from a cross between emmer and a second species of weedy goatgrass. At the time of writing, the date that this occurred is still very uncertain. According to one study, it may have happened as recently as 8,000 years ago somewhere in a cultivated field in the Fertile Crescent and after emmer had been domesticated, but another study puts the date back to at least 230,000 years ago and so before the origin of modern humans. Regardless of exactly when these two hybridizations in the evolutionary history of bread wheat actually occurred, each of them added a whole extra set of chromosomes, so that the species now has three complete sets.

This giant genome, five times bigger than our own, gives bread wheat enormous genetic potential for evolution. The reason that it does so is that natural and artificial selection require genetic variation as a raw material from which to fashion new forms. The ultimate source of genetic variation is mutation or random mistakes that mostly occur when DNA is copied. As you might expect, random mutation more often than not does damage. In an organism with only one set of chromosomes, the damage caused by mutation can slow the rate of evolutionary change, but when there are three sets, there is room to experiment. Metaphorically speaking, bread wheat has a belt, suspenders, and elastic all holding up its genetic pants. This triple-strength genome accounts for the enormous evolutionary versatility of bread wheat, expressed in the many, many varieties that are adapted to different environments.

Genetic variation is the raw material of evolution, and the genetic variation contained in different local populations is the raw material used by breeders to improve crops. Locally adapted varieties of crops are like the local versions of a language, containing novel words (novel genes) that can find a use outside their place of origin. English is full of such words, particularly for food and drink. "Whisky," for example, comes from the Gaelic, "chocolate" from Nahuatl, "chutney" from Hindi, "bagel" from Yiddish, "hominy" and "persimmon" from Powhatan. Local crop varieties, called landraces, are just as individual as dialects because in addition to artificial selection to suit the local tastes of their growers, they have experienced thousands of years of natural selection adapting them to the local climate and conferring resistance to endemic diseases. Such adaptation is a matter of life-and-death, not just for crops, but for us too.

Cereal domestication vastly increased the amount of food available for the human species, but it also made us dependent upon crop health for survival. The ancient Egyptians, so dependent upon bread, knew famine. In the Old Testament book of Genesis, Pharaoh has a strange dream:

> And he slept and dreamed the second time: and, behold, seven ears of corn came up upon one stalk, rank and good.
>
> And, behold, seven thin ears and blasted with the east wind sprung up after them.
>
> And the seven thin ears devoured the seven rank and full ears. And Pharaoh awoke, and, behold, it was a dream.

According to the biblical story, Pharaoh asks his soothsayers what the dream means, but (most improbably) they are stumped for a reply and so he sends for the Hebrew servant Joseph, who has a reputation for successfully interpreting dreams. Joseph says:

> Behold, there come seven years of great plenty throughout all the land of Egypt: And there shall arise after them seven years of famine; and all the plenty shall be forgotten in the land of Egypt; and the famine shall consume the land.

Joseph then sagely recommends that Pharaoh store grain produced in the years of plenty to prevent famine in the years when crops will be poor. Excellent advice.

Of course, cereal grains and other seeds store so well precisely because this is the function in the plant life cycle that natural selection has designed them for. Seeds are plants' infant food stocks that we pirate for our own use. This makes us parasites on our crops, but unfortunately there are other parasites too. We are in competition with viruses, bacteria, fungi, rodents, and insects, like the locusts that figured among the ten plagues in the book of Exodus.

Diseases caused by rust fungi are some of the greatest threats to cereal crops because these fungi have short life cycles that enable them to evolve quickly, and they are very easily spread by tiny windborne spores. A strain of stem rust called Ug99 that emerged in Uganda in 1998 spread very quickly throughout the wheat-growing regions of Africa and threatens over a third of the world's wheat crop. Ninety percent of bread wheat varieties are susceptible to Ug99, but fortunately it has been possible to breed high-yielding varieties with resistance to Ug99 using genes from the rarer resistant varieties.

Food security for all our crops depends on being able to continually match the challenge posed by constantly evolving diseases. The armory in this fight is the collection of crop varieties and landraces containing genes that protect against disease. Arguably, the plant breeder who made the biggest contribution to stocking the global granary was the Russian scientist Nikolai Ivanovich Vavilov (1887–1943). His is the tragic story of a scientist, now acknowledged as a Russian national hero responsible for helping to feed tens of millions, but who died of starvation in a Soviet prison.

After graduating from agricultural college, Vavilov started research into crop diseases with the aim of alleviating the famines that periodically ravaged Russia. He realized that the differences in disease resistance to be found between one crop variety and another could be understood through the then-new science of genetics, and in 1913 he took the opportunity to go to Cambridge, England, to study with one of its founders, William Bateson.

While in Cambridge, Vavilov found inspiration for his future research in reading books from Charles Darwin's personal library, which was then kept in the university. He was particularly impressed by Darwin's evident interest in hereditary variation in crops and the role of geographical variation in the evolution of new species. On returning to Russia at the outbreak of the First World War, Vavilov began three decades of tireless collecting, research, and travel.

Vavilov's collecting took him on long expeditions to Europe, North Africa, North and South America, the Caribbean, Afghanistan, China, Japan, and southwest Asia, gathering seeds of crop plants wherever he went (map 4). He sent hundreds of kilos back to his institute in Leningrad whenever he could, including notes on disease resistance and a record of the altitude and location where each sample was taken. By the early 1930s, he had assembled a collection of no fewer than 200,000 samples, including 30,000 varieties of wheat that were cultivated near his institute in Leningrad.

His collecting journeys were guided by his theory that the greatest genetic diversity for any crop would be found in the area where it was first domesticated. This theory has not stood the test of time, but it did cause him to discover that some of the highest genetic diversity is to be found in mountainous regions. This led him to explore some of the most inaccessible and dangerous places.

In the late 1930s, Vavilov began writing a book to be called *Five Continents* that would tell the tale of his plant-collecting adventures, but its publication was overtaken by the Stalinist purges that cost many scientists working with him, and finally Vavilov himself, their lives. For twenty years the manuscript was believed lost, but in the early 1960s, by which time Vavilov had been posthumously rehabilitated, his secretary A. S. Mishina revealed that she had courageously hidden a large part of the book from the secret police.

One of the areas that Vavilov had desperately wanted to visit was Abyssinia (modern Ethiopia) and the mountains of the adjacent Eritrea. In *Five Continents*, Vavilov describes how he met with Ras Tafari—who later became Emperor Haile Selassie, venerated by the Rastafarians—to obtain permission to travel through Abyssinia. The Ethiopian highlands did not disappoint in their diversity, and Vavilov wrote: "The fields display such an incredible mixture of varieties. It was necessary to collect hundreds of seed heads just to obtain a representative sample of the botanical composition of a single field." In a field near Aksum, on the banks of the upper Blue Nile, Vavilov wrote with great excitement that he discovered a type of durum wheat that plant breeders had

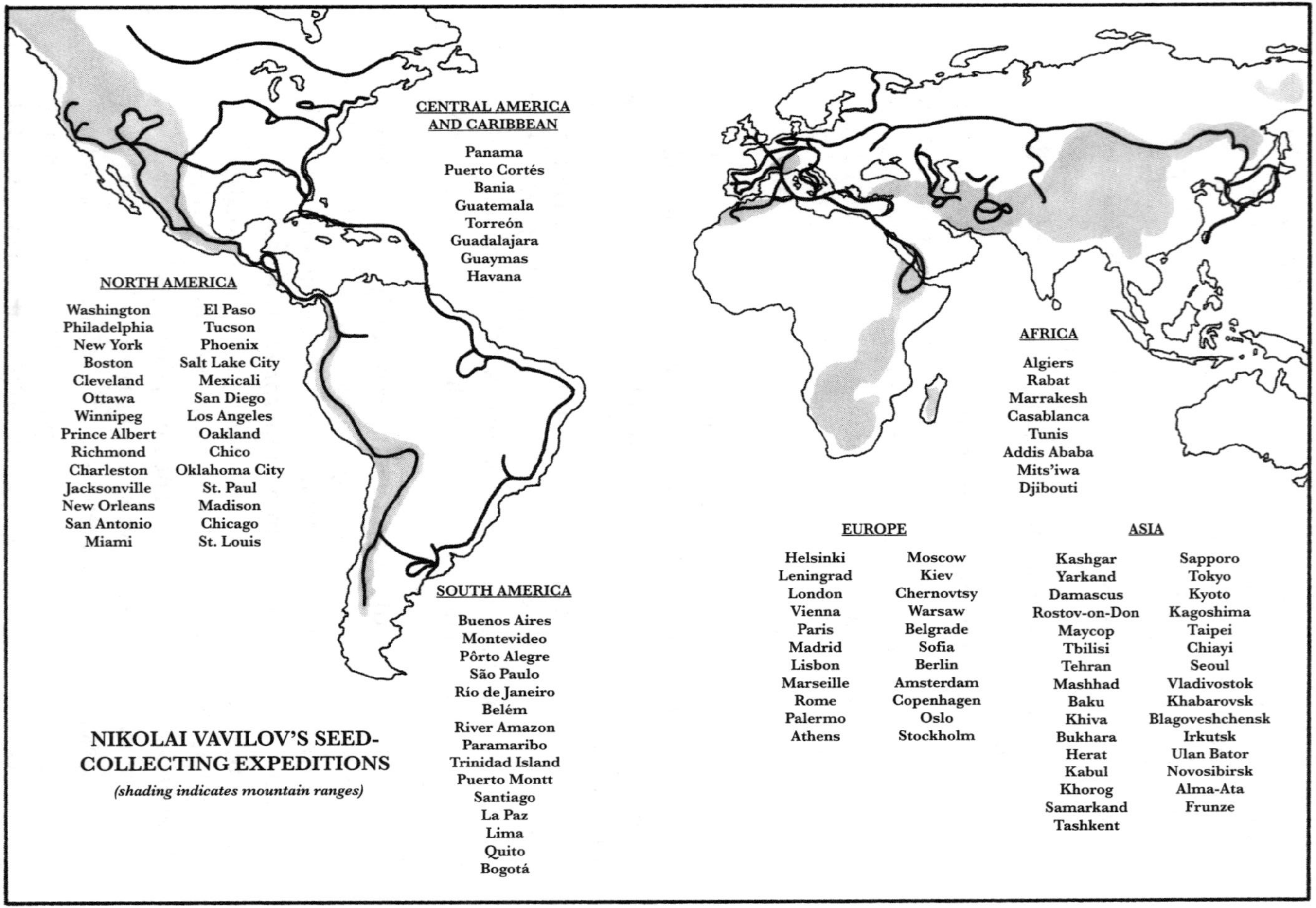

NORTH AMERICA

Washington
Philadelphia
New York
Boston
Cleveland
Ottawa
Winnipeg
Prince Albert
Richmond
Charleston
Jacksonville
New Orleans
San Antonio
Miami

El Paso
Tucson
Phoenix
Salt Lake City
Mexicali
San Diego
Los Angeles
Oakland
Chico
Oklahoma City
St. Paul
Madison
Chicago
St. Louis

CENTRAL AMERICA
AND CARIBBEAN

Panama
Puerto Cortés
Bania
Guatemala
Torreón
Guadalajara
Guaymas
Havana

SOUTH AMERICA

Buenos Aires
Montevideo
Pôrto Alegre
São Paulo
Rio de Janeiro
Belém
River Amazon
Paramaribo
Trinidad Island
Puerto Montt
Santiago
La Paz
Lima
Quito
Bogotá

NIKOLAI VAVILOV'S SEED-
COLLECTING EXPEDITIONS

(shading indicates mountain ranges)

EUROPE

Helsinki
Leningrad
London
Vienna
Paris
Madrid
Lisbon
Marseille
Rome
Palermo
Athens

Moscow
Kiev
Chernovtsy
Warsaw
Belgrade
Sofia
Berlin
Amsterdam
Copenhagen
Oslo
Stockholm

AFRICA

Algiers
Rabat
Marrakesh
Casablanca
Tunis
Addis Ababa
Mits'iwa
Djibouti

ASIA

Kashgar
Yarkand
Damascus
Rostov-on-Don
Maycop
Tbilisi
Tehran
Mashhad
Baku
Khiva
Bukhara
Herat
Kabul
Khorog
Samarkand
Tashkent

Sapporo
Tokyo
Kyoto
Kagoshima
Taipei
Chiayi
Seoul
Vladivostok
Khabarovsk
Blagoveshchensk
Irkutsk
Ulan Bator
Novosibirsk
Alma-Ata
Frunze

Map 4

futilely been trying to produce for decades, while here nature herself had produced just such a plant.

Farther down the road toward Eritrea, Vavilov's companions became agitated, fearing bandits. "To encourage them it was necessary for me to go in front. We had succeeded in advancing for only a few hours after crossing the river when people with guns, obviously used to attacking caravans, appeared from behind a dense thicket." Seeing a European at the head of the caravan and knowing that Europeans traveled heavily armed, the bandits began polite bowing and invited the expedition to spend the night in their village. "It was late and we had to stay somewhere for the night, but how should we deal with this?" The Russians decided to load their best revolvers, to drink enough wild coffee to ensure that they would not doze off during the night, and to send the leader of the gang their last two bottles of five-star brandy as a gift. "The guide returned from the mission slightly tipsy but with fried chicken, a pot of honey and armfuls of flat cakes made of teff."

Teff is a grain that has tiny seeds, much smaller than those of wheat, rye, or barley but, like those domesticates, its seed heads have been artificially selected so that they do not shatter. Although teff is endemic as a crop to Ethiopia, the wild and weedy relative from which it was domesticated is remarkably widespread in tropical and temperate regions. Only in Ethiopia did people develop a taste for the bread made from this unusual flour and turn the weed into a crop. Teff seeds lack gluten, and so dough made from it does not have the elastic properties that are essential to the leavening of wheat bread. Instead, teff flour is mixed with water and spices and left to ferment and thicken. The resulting batter is then poured onto a hot griddle pan to make huge pancakes called injera. These are moist, springy, and pitted with tiny holes left by the escape of gas during cooking. Injera has a slightly sour taste, and like other flat breads the world over, it is used at the table to scoop up or wrap other foods into morsels that are popped into the mouth.

Despite the botanically unique gift of injera and the delicious accompaniments that Vavilov's guide brought back from the bandit camp, the Russian wisely decided that the gang's hospitality was not to be trusted. Alcohol and coffee, selectively administered to sedate the thieves and rouse their intended victims, allowed Vavilov's party to effect their escape. At 3 a.m. the expedition packed up and hastily departed, leaving the bandits still sleeping it off.

Vavilov's life was brutally cut short by an ending that was as tragically ironic as his career was heroic. The scientist who survived the hardships of travel in the remotest corners of the world, all in the cause of ending famine in his homeland, was falsely accused by Stalin's secret police of treachery and sabotage. In 1940 he was imprisoned, tortured, and slowly, deliberately killed by starvation in the very town where he began his professional career.

But there is a bittersweet coda to Vavilov's life. In June 1941 the German army rapidly advanced across the borders of the Soviet Union, and by September they reached the gates of Leningrad, where they were halted by fierce resistance. Despite the persecution of Vavilov and his staff, the Soviet authorities recognized that the seed collections held at the institute in the city needed to be rescued, and plans were made to evacuate them. The Germans also had plans to seize the seed collection and had formed a special unit of the SS called the Russland-Sammelcommando to do so. A small part of the collection was successfully moved to safety by the Russians, but the largest and most important part remained in the besieged city, along with a nucleus of dedicated staff who stayed to protect it. Many of these scientists died of starvation while guarding the precious seeds that could have assuaged their hunger.

The Germans' bombardment was intended to flatten Leningrad, but Hitler's hubris inadvertently protected the institute and its collections from utter destruction. The Nazi leader was so confident that he would take the city that he had already ordered the invitations to be printed for a victory party that he planned to hold at the Astoria Hotel. By chance, Vavilov's institute was near the Astoria and the German consulate and so received protection from the worst of the shelling.

The true value of Vavilov's legacy was finally recognized only in 1979, when his biographer G. A. Golubev assessed the impact of his seed collections and breeding programs on Soviet agriculture. He calculated that 80 percent of the cultivated land in the Soviet Union was sown with crop varieties derived from the collections of Vavilov's institute. A thousand new varieties carried the name Vavilov, and these contributed 5 million tons of extra produce per year, worth more than $1.5 billion at the official exchange rate of the time.

The genetic diversity held in seed collections like the one that Vavilov created enables crops to be grown in a broader range of climates and over a wider geographical area than the wild ancestors of the same crops could tolerate. Bread wheat, with its accommodating genome and hundreds of thousands of varieties, is a good example. However, even wheat has its limits, and global yields of the crop are already being negatively affected by global warming of the climate. Where the potential of varietal diversity has pushed a crop to its climatic limits, the best strategy may be to switch to another crop species that is better adapted to the prevailing climate. Vavilov saw just such a change happening on his first expedition.

In 1916, a year before the Russian Revolution, Vavilov was on a collecting expedition in Persia (Iran), collecting landraces of barley, rye, and wheat, including a local variety of the latter that was entirely immune to mildew. While making these collections, Vavilov observed that winter wheat fields were badly infested with weedy rye and that as he proceeded higher in the mountains, rye increasingly became the crop in place of poorly performing wheat. From this discovery he formulated the idea, now generally accepted, that rye began as a weed of wheat fields and was accidentally domesticated through being harvested with it and being used as a substitute where wheat grew poorly.

Rye is a much hardier crop than bread wheat, better suited to poor soils and cold climates, and can be grown as far north as the Arctic Circle. The grain has a high protein content and contains unusual carbohydrates called arabinoxylans (or pentosans). Arabinoxylans are able to absorb a lot of water, which in nature aids germination of the rye seed, and in cooking gives rye flour four times the water-holding capacity of wheat flour. Wheat bread stales quickly because the starch in it crystallizes and hardens after cooking and cooling, though this process is reversible, which is why warming freshens it. By contrast, arabinoxylans remain soft when cooled, giving rye bread a much longer shelf life.

Rye was the bread of the poor in northern and eastern Europe, and is still popular there. In the nineteenth century, immigrants from these areas created a demand for rye in the United States, and the crop was widely grown there until the 1960s. Then as the demand for rye diminished and less and less of the crop was grown, something quite strange happened—rye began appearing as a weed in other crops. By the beginning of the twenty-first century, weedy rye infested a million acres of cropland in the western United States, causing losses of $26 million a year. Various theories were advanced as to what had happened; perhaps this was a new hybrid, or maybe the crop had become self-sustaining in the fields where it was once deliberately sown? Studies of the characteristics of weedy rye and of its genetics showed that both these ideas were wrong. What had happened was that, having been accidentally domesticated in the Old World and then carried from there to the New World as a crop, rye had evolved back into a weed in North America. A change in a single gene had restored seed shattering, which is of huge advantage to dispersal in a wild plant, and seeds had also become smaller like those of wild rye. There is no better demonstration than this of how evolution is constantly at work, unless it is how agriculture has shaped our own recent evolution.

The process of crop domestication wrought major evolutionary changes not only in crop plants, but also, by direct and indirect means, in our own species too. Indeed, the changes in human society were so momentous that the Australian historian V. Gordon Childe, writing in the 1930s, described the event that occurred in the Neolithic, some 10,000–12,000 years ago, as a revolution. The importance of the Neolithic Revolution is difficult to exaggerate. The need to tend fields made settlements permanent, and the surpluses of food produced by agriculture allowed populations to grow and released labor for occupations quite unrelated to the basic necessity of food gathering. That other turn-ing point in human history, the Industrial Revolution, could not have occurred without the Neolithic Revolution that preceded it by more than 10,000 years.

Though agriculture produced an abundance of food, the change to a cereal-based diet rich in starch was not an especially healthy one for the first farmers in the Fertile Crescent. There is evidence of how we adapted to this radically new diet in our saliva. It is hardly polite to drool over the prospect of food, but delicious smells are more delicately described as "mouthwatering" precisely because the salivary glands are stimulated by food odors into pumping out saliva in anticipation of a meal. Saliva is mostly water, but it also contains a variety of different enzymes including some that begin the process of digestion, which starts in the mouth, not the stomach. Up to half the protein content of saliva may consist of an enzyme called α-amylase that breaks down starch into sug-ars, but not everyone has the same amount of α-amylase in their saliva.

The amount of α-amylase in your saliva can be influenced by a variety of things such as stress, but a major cause of the varia-tion between one person and another is the number of copies of the α-amylase gene that they have inherited—there can be any-where between one and fifteen copies. Why exactly there should be such variation in the number of copies of this particular gene is not at all clear, but the Neolithic Revolution seems to have raised the average number of copies in populations that eat a lot of starch.

A study compared the number of copies of the α-amylase gene in three populations with high-starch diets with the number in four populations with a low-starch diet. In the high-starch group were Japanese and Americans of European descent who tradi-tionally eat a lot of cereals (rice, wheat, and corn) plus an African hunter-gatherer tribe called the Hadza, who do not practice agri-culture, but who gather and eat starchy roots and tubers. In the low-starch group were people from three other African tribes and one from Siberia. The study found that people in the high-starch group had on average about two more copies of the α-amylase gene than did people in the low-starch group. This suggested that the extra number of α-amylase genes might be an evolutionary adaptation to a high-starch diet.

It is easy to imagine how natural selection would have acted on the preexisting variation in gene copy number that must have been present in pre-farming communities, favoring individuals who could better digest starch when they began to eat bread and rice, or roots and tubers. A fly in the ointment for this theory is that actually most starch digestion does not occur in the mouth but in the stomach, where another amylase enzyme occurs that is secreted by the pancreas. In contrast to the situation with salivary α-amylase, the gene for pancreatic amylase has not been duplicated and so does not vary in the number of copies carried by different people. Nonetheless, the salivary amylase that is mixed with food in the mouth does continue to work when it reaches the stomach, so maybe people with more α-amylase genes really do digest starchy food more efficiently than those with fewer copies of this gene. This efficiency hypothesis can easily be tested.

When starch is fully broken down, it yields the sugar glucose, which is the molecule that fuels all living cells. Therefore, if the efficiency hypothesis is correct, after eating starch, someone with many copies of the α-amylase gene ought to have more glucose turn up in their bloodstream than someone with only a few copies of the gene. Surprisingly, when this experiment was carried out, quite the reverse happened—people with lots of salivary amylase had significantly less glucose in their blood than people with little of the salivary enzyme. What could be going on?

The amount of glucose in the blood is finely regulated by the hormone insulin. Having too much fuel circulating around in the bloodstream is just as dangerous for the body as having too much gasoline sloshing about an engine would be for a vehicle. It appears that people on a high-starch diet with high numbers of α-amylase genes do benefit from this genetic constitution. However, the benefit is not that they digest starch more efficiently, but rather that their bloodstream is not dangerously flooded with glucose after a starchy meal. Since too much glucose in the blood can lead to type-2 diabetes, this is certainly an advantage that natural selection would take notice of. If this hypothesis is correct, then the function of salivary amylase is not merely to begin the digestion of starch, but to release sugars from it into the mouth, sending an early warning via taste receptors that a lot of starch is on the way to the stomach. Insulin can then be released in anticipation, and dangerously high blood glucose levels can be prevented.

Cereal domestication did not just select for genetic changes in our own ability to live on a high-starch diet—but it affected the evolution of humans' best friend too. Dogs were domesticated from wolves by at least 10,000 years ago, probably earlier, so they have been feeding at our table or from our leftovers since the dawn of agriculture. Dogs do not have salivary amylase as humans do, but a comparison between the genomes of dogs and wolves, which are their wild ancestors, shows that three genes affecting the digestion of starch changed during domestication. One of these changes was a big increase in the number of copies of the gene that supplies the enzyme amylase to the digestive system of dogs. Evolution has adapted dogs to thrive on the starchy crumbs from our table.

Our daily bread, the quotidian food that most of us take for granted, has a hidden 12,000-year history that changed us in every way imaginable. It was the cornerstone of the Neolithic Revolution when we learned to direct the evolution of plants and animals for our own ends. Agriculture fed and amplified the human population, producing surpluses that could be used to build cities for the living and grand tombs for the dead. Agriculture also gave us the leisure to contemplate nature and ultimately to discover its laws. Looking at the effects of domestication on animals and plants, Darwin saw that the process of artificial selection that molded these organisms to our taste was analogous to natural selection that fashioned us and every other living thing. Bread domesticated us, and now we are completing the domestication of the planet.

With a morsel of bread, we have sailed back to the dawn of agriculture and then forward to the evolutionary consequences that domesticating crops have had on our species. The aroma of fresh bread has stimulated our gastric juices, and the starch in our mouths has physiologically prepared us for what is to come. I think it's time for the soup, don't you?

# 5

## *Soup—Taste*

Soup is a reminder that everything important in life boils down to substances dissolved or suspended in water. Life itself began that way in the sea, quite possibly around deep-sea hydrothermal vents that gush hot water and cook the seafloor to a temperature at which interesting chemistry begins to occur. Charles Darwin avoided speculating in print about the origin of life, but in a letter written in 1871 to his friend the botanist Joseph Hooker, he allowed himself to imagine that it might have begun "in some warm little pond with all sorts of ammonia and phosphoric salts—light, heat, electricity &c. present."

The evolutionary biologist and polymath J. B. S. Haldane (1892–1964) later called this the "primordial soup," and that label has firmly stuck. The proponents of rival theories of the origin of life have occasionally signaled their dissent by proposing that life began in a primordial crêpe or even a primordial vinaigrette, but soup is the perennially favorite starter on life's menu. So much so, in fact, that a Swiss food scientist has even proposed that all of the early steps in the transition from a non-living primordial soup to life itself can be achieved with kitchen chemistry, starting with just polysaccharides such as starch. Personally, while I'd agree that potato soup can be life-giving, I'd hate to have to cook it on a hydrothermal vent.

Jean Anthelme Brillat-Savarin (1755–1826), the celebrated French author of *The Physiology of Taste*, claimed that nowhere was soup better than in France, and no wonder, he wrote, because "soup is the basis of our national diet, and centuries of experience have brought it to its present perfection." The Mock Turtle, in Lewis Carroll's *Alice in Wonderland*, was equally enthusiastic:

> Beautiful Soup, so rich and green,
> Waiting in a hot tureen!
> Who for such dainties would not stoop?
> Soup of the evening, beautiful Soup!
> Soup of the evening, beautiful Soup!

Harold McGee, in his indispensable book *On Food and Cooking*, invites us to gaze into the center of a bowl of hot Japanese miso soup and to watch as the convection currents cause fluffy particles to rise in billowing clouds, as if we are gazing godlike from the sky on to the world below. Soup, it is clear, can be full of wonder. It is also full of taste. Our sense of taste tells us when the juices in our mouth contain substances that are nutritious or that might be poisonous. Five types of sensor cells on the tongue distinguish the tastes salt, sweet, sour, bitter, and umami. A growing number of scientists believe that the taste of fat, for which there are also sensors on the tongue, is a sixth taste. Aristotle thought so too.

Miso soup tastes salty, but it also has the delicious, mouthfilling savory taste called umami. While salt, sweet, sour, and bitter have been recognized as distinct tastes for millennia, umami was not identified until 1909. In that year Kikunae Ikeda, a chemistry professor at the Imperial University in Tokyo, published a paper in Japanese in which he said that he thought there was at least one more taste than the four then commonly recognized: "It is the peculiar taste which we feel as *umai* [savory], arising from fish, meat and so forth. The taste is most characteristic of broth [*dashi*] prepared from dried bonito and seaweed [*Saccharina japonica*]. While it is based on a subjective sensation, many people who are asked always agree to this conjecture either immediately, or after brief consideration. . . . I propose to call this taste *umami*." The suffix *mi* indicates "essence" in Japanese, so *umami* means "savory essence."

While Ikeda was convinced that umami was a distinct taste that had been lurking unrecognized under our very noses, he had to identify its chemical basis in order to prove its existence. He knew that whatever the compound was, it must be water soluble and present in seaweed, so he began his chemical analysis with an aqueous seaweed extract, or as any cook would call it, seaweed soup. There followed a laborious process of evaporation, distillation, crystallization, precipitation, and seemingly every -*ation* known to the nation, making thirty-eight steps in all. Ikeda finally ended up with some gritty-tasting crystals that tasted of seaweed broth. A little bit more chemical wizardry, and Ikeda was able to show that the purified crystals were in fact glutamic acid. The sodium salt of glutamic acid, called sodium glutamate, proved to give the best umami taste.

Ikeda modestly stated that "this study has discovered two facts: one is that the broth of seaweed contains glutamate and the other that glutamate causes the taste sensation *umami*." What he had actually achieved was something much more important: he had identified a fifth taste in our gustatory repertoire. Not only that, but Ikeda made two other important contributions, one of them theoretical and the other practical. On the theoretical side, he considered why we have a taste for umami. Glutamic acid is present in many protein-rich foods such as meat, and since it can be tasted even in trace amounts, it provides a very good signal that the food we are tasting is nutritious. Human breast milk contains glutamate in ten times the concentration found in cow's milk. The pleasure we derive from tasting umami looks as though it is natural selection's way of making sure that we eat the right things.

On the practical side, Ikeda patented a method for producing monosodium glutamate (MSG), which is now very widely used as a flavor enhancer in cooking. Sodium glutamate comprises as much as 3 percent of the dry weight of seaweeds such as *konbu* (*Saccharina japonica*). Annually, 2.5 billion tons of this species are harvested in China. There is a biological reason why seaweeds are such a good source of MSG. Every cell has a membrane around it that encloses and protects the cell contents. Cell membranes are semi-permeable and allow small molecules like water to pass in and out. When two solutions of different concentration are separated by a semi-permeable membrane, a process called osmosis takes place in which water molecules move from the less concentrated solution to the more concentrated one. Osmosis only stops when the salt concentrations on either side of the membrane have been equalized by the movement of water across it. Fresh seaweeds are up to 90 percent water, so imagine what would happen to their cells when immersed in seawater if osmosis got to work—the high concentration of common salt in seawater would cause them to quickly lose water, shrivel up, and die. The solution is in the solution. Sodium glutamate in seaweed cells helps to equalize the difference in salt concentration between the sea and the weed, thus preventing dehydration and collapse. As you might expect, seaweeds from the most saline oceans contain the greatest concentrations of MSG.

If you prefer to use a less industrial source of glutamate than the white crystalline substance extracted from seaweed, you will also find it naturally present in cooked tomatoes and various fermented foods such as soybean paste and miso. The grittiness of a good parmesan cheese is due to crystals of MSG that form naturally during the aging process. Sprinkle a little on your minestrone soup!

Not long after Ikeda's discovery that glutamate extracted from seaweed tasted umami, one of his students isolated a molecule called inosinate from dried bonito, the other main component of *dashi*, that also tasted umami. Inosinate is a ribonucleotide—the class of compounds that put the "N" in DNA (deoxyribonucleic acid)—and hence also important nutritionally. So, *dashi* contains a double dose of umami-tasting substances. Many decades later in the 1950s, a Japanese food scientist studying yeast found that when broken down, it released a ribonucleotide called guanylate that also had an umami taste. He further discovered that mixing either of the nucleotides guanylate or inosinate with glutamate increased the umami taste hugely over the taste that any of the molecules could elicit alone. Here then, in simple chemistry, was an explanation of why *dashi* is such a good soup base: it combines two molecules, glutamate from seaweed and inosinate from dried bonito, that synergistically trigger an umami taste bomb.

*Dashi* is perhaps the purest traditional source of umami in a liquid medium, but a good stock is a fundamental starting point for any soup, and virtually all recipes involve obtaining an umami-rich solution by gently simmering a source of protein such as bones or fish bits. Chicken stock is such an excellent source of glutamate that some cuisines rely on almost this alone as a soup base. The animal ingredients in stock are the predominant source of glutamate, and the nucleic acid key to umami taste can be supplied by inosinate from the same source or by guanylate when plants or mushrooms are added to the pot.

A vegan version of *dashi* may be made by substituting the fish stock with dried shiitake mushrooms. Shiitake and, indeed, many other edible fungi become rich sources of guanylate and glutamate when dried and then rehydrated in tepid water. Rehydration should not be done in hot water as this will destroy the mushroom's enzymes that release the molecules that give it flavor. Cooked tomatoes have many virtues as a component of sauces and soups; one of them is that the glutamate they contain helps unlock umami. Mushroom and tomato pizza, anyone?

Although umami was right under our noses all along, it was many decades before its existence as a fifth taste was accepted outside Japan. One reason this recognition may have been slow in coming is that common salt (sodium chloride) and sodium glutamate taste somewhat similar, so it could be argued that the taste of sodium glutamate was just the taste of salt. However, anyone able to read Ikeda's paper would have found a ready answer to this problem. Ikeda pointed out that the taste of common salt becomes undetectable at concentrations lower than 1 part salt to 400 parts water, while sodium glutamate can still be tasted right down to a dilution of 1:3,000. He further comments that tests of the quality of soy sauce, which is rich in umami and also contains a lot of salt, are based on exactly this principle. A good soy sauce is one that retains taste after that of the salt content has been diluted away. Nearly a century passed between the publication of Ikeda's paper in 1909 and the appearance of biological evidence that clinched the truth of his discovery.

All our experiences of the world, including what edible parts of it taste like, are acquired through a cascade of events that starts with specialized cells in our sense organs and are transmitted via nerve pathways to the brain. The sense organs for taste are taste buds found on the upper surface of the tongue and on the palate. Ikeda found that tasting glutamate signaled "umami!" to his brain, but experience is subjective and others tasting glutamate thought that they just got the signal "salty." Even the dilution test that showed that glutamate could be tasted at lower concentrations than salt did not convince skeptics that the two tastes were quite different.

Final confirmation that umami really is a distinct taste came with the discovery in the early years of the twenty-first century that there are cells in taste buds that are furnished with proteins on their outer surface that react specifically to glutamate and guanylate or inosinate, but not to salt. These proteins belong to a family of molecules called receptors that behave like tiny locks on the gateway to taste. Only molecules of the right shape and chemistry can unlock the receptor, which then triggers a signal to the brain that says "umami." Of course, we may not consciously register the umami taste, but just think "mmm, that's nice."

It turned out that the umami receptor actually consists of not just one, but a pair of receptor proteins, which explains why it reacts much more strongly when two different keys are applied to the lock rather than just one. The first key is glutamate, but the second can be either of the two nucleic acids: guanylate found mostly in cooked plants and fungi, or inosinate from ingredients of animal origin. These nucleic acids are released when cells are broken down during cooking, decomposition, or fermentation. The combination of glutamate plus a nucleic acid is a better indicator of the nutritional quality of food than is glutamate on its own.

The two proteins in the umami receptor are produced by a pair of genes called *T1R1* and *T1R3* (*Note:* Names of genes are italicized, e.g., *T1R1*, while the proteins they encode are written in roman, T1R1). Ever parsimonious, evolution has fashioned the receptor molecule for sweet-tasting substances like sugar from a combination of the T1R1 protein with another called T1R2. This family of three similar taste receptor proteins involved in sensing two important types of nutrient may have a common evolutionary origin in a single ancestral gene, though at the time of writing, this hypothesis is untested.

People tend to think of evolution as a directional process, like a car that has no reverse gear, but this is not how it works at all. The traits that natural selection has winnowed from random mixtures of the useful and the useless can be undone if they are no longer advantageous. Genes for traits that have lost their function during evolution tend to accumulate mutations and to become ghost-like "pseudogenes" that are but pale shadows of their once-useful selves. Thus in cats and certain other carnivores that eat only meat, the ability to taste sugar is redundant and the gene for the T1R2 protein no longer functions. Your cat has no taste for sugar, even if presented as delectable sugar mice. Bears are carnivores but also eat berries, and so they still have the *T1R2* gene that is essential to the sweet taste. The giant panda is a relative of the bears and eats only bamboo so, as one might expect, it can taste sugars but not umami and the *T1R3* gene is non-functional. Sea lions swallow their food whole, without chewing it first, so in these fish-eaters both the umami and sweet receptors are redundant and all three of the *T1R* family have become pseudogenes. The same evolutionary loss has occurred independently in dolphins and vampire bats, which don't chew either. Perhaps there is a cautionary tale here for children of the genomic age: "You chews or you lose."

While I have distracted you from cooking with tales of cats, bears, and sea lions, our soup is bubbling nicely, so let's taste it. It has a satisfying fullness that we recognize as umami, there is a slight zing from the dash of wine vinegar we added, but something is missing. What is it? Of course, it needs a pinch of salt! Like each of the other four basic tastes, there are taste receptor cells dedicated to sensing salt. Salt, or sodium chloride (NaCl), when in solution dissociates into a positively charged sodium ion $Na^+$ and a negatively charged chloride ion $Cl^-$. It's the sodium ion that we taste, or even crave, and this enters the specialized salt receptor cells through channels in its outer membrane.

Sodium is essential to animal life, and it is an important constituent of all the fluids in the body, where its concentration is finely regulated. Low concentrations of salt are pleasant and can improve flavors, even when present at well below concentrations that taste salty, but high concentrations of salt, on the other hand, can be aversive. No one willingly drinks seawater. There are nightclubs that salt the water in the faucets of their bathrooms to force thirsty clubbers to buy expensive bottled water with which to rehydrate.

Studies in mice have found that there are actually two types of salt-sensing taste receptor cells. One type senses low concentrations of sodium (and only sodium), and this stimulates attraction to salt. The other type can detect only high concentrations of sodium chloride as well as other salts. Stimulation of this second type of salt taste receptor cell leads to salt-avoiding behavior. Whether we humans also have two types of taste receptor cells for salt is not known, but it seems very likely that we do. If so, then it would be logical to think of "nicely salty" and "nastily salty" as separate tastes, making at least six basic tastes in all. I doubt that I am the only person who has been served soup that tasted unpleasantly salty.

Sugar, salt, and umami are the nice tastes, but they have a couple of ugly sisters who are bitter and sour. Bitter is the taste that makes us involuntarily pull a face, and if you think of the foods that taste bitter, you will realize that they are all derived from plants. All the greens in the cabbage family such as Brussels sprouts, cabbage, kale, and broccoli are inherently bitter, even though their bite has been tamed by domestication. The bitterness is untamed in watercress and arugula (salad rocket) and possibly even enhanced by domestication in mustard and its relatives wasabi and horseradish, which we, seemingly perversely, enjoy.

The bitter taste of mustard and its relatives comes from a class of compounds called glucosinolates. They are defensive molecules that deter nibbling insects, though there are specialist caterpillars that can eat them, as you will know only too well if you grow your own vegetables. In fact, the scientific name of arugula is *Eruca*, the Latin word for caterpillar. Even though not successful at protecting brassicas from all caterpillars, glucosinolates can protect plants against fungal diseases such as mildew.

There is no poison that, through long association, natural selection has not equipped some animal to tolerate, though it is usually achieved at the cost of the animal being very specialized in its diet. Cucurbitacin is the chemical that makes cucumbers, squash, and their kin bitter. The omnivorous two-spotted mite is poisoned by cucurbitacin, but the cucumber beetle can tolerate this substance, the smell of which attracts it like a dinner gong to cucumber plants.

You might think that bitter greens would make poor soup, but such is the magic of cooking and the complexity of our reactions to taste that watercress excels in soup made with cream or potato, or in a Chinese broth made with pork ribs. Mustard also adds flavor to thick soups such as cream of onion, gammon, Gruyère cheese, Stilton, and almond. Beyond the soup course, arugula is classically paired in a salad with shaved parmesan, hitting bitter, salt, fat, and umami tastes all in one go.

Another large class of bitter plant compounds are the flavonoids, largely absent from soups but enjoyed in tea, where its bitter edge can be dulled with lemon or milk. Yet another group of bitter-tasting compounds that plants use to defend themselves are the alkaloids. These include deadly poisons such as strychnine and psychoactive drugs such as morphine, cocaine, and caffeine. Think how bitter coffee can be. Quinine is notoriously bitter, but we enjoy the taste that just a small trace of it imparts when sweetened in Indian tonic water. The bitterness of unsweetened chocolate is not to everyone's taste, but the alkaloid called theobromine is essential to the flavor and enjoyment of chocolate.

The curious thing about the bitter taste is just how many different compounds there are that trigger this one sensation. There are only a few dozen molecules that taste sweet, fewer than a handful that taste umami, but thousands that taste bitter. The reason for this is that most plants defend themselves with poisons of some kind, so plant-eating animals are equipped by evolution to detect these. There is only one kind of bitter taste cell in our taste buds, but it has up to 25 different types of receptor protein on its surface, each one produced by its own *TAS2R* gene. Using the lock-and-key analogy employed earlier, there are 25 different locks on the cell that triggers bitter taste, activation of any of which will set off the bitter alarm signal to the brain. The more types of key (molecule) that can unlock the bitter response, the more effective the alarm system is and the better protected we are. Some of the receptors are finely tuned to detect just one bitter compound, but it seems that most have a broader sensitivity and respond to many, and may overlap in the bitter compounds they detect. Three different receptors sense the bitter taste of hops in beer, for example.

We share the genes that give us this wide response to bitter tastes with mice and other mammals. Our ancestors and those of mice parted company 93 million years ago, so the taste genes that we have in common have deep roots in our evolutionary past. Animals with vegetarian habits have many more receptor genes for bitter compounds than animals that don't eat their greens. Cats have only 6, but mice have 35. Our own repertoire of 25 bitter receptor genes suggests that our ancestors ate a wide variety of plants, just as our cousins the great apes do today. The 11 pseudogenes found in the human genome that once coded for bitter receptors are ghosts from this time long ago.

An ingenious experiment illustrates the point that the locks (receptor proteins) do the detecting of the bitter-tasting or sweet-tasting molecules in our food, but whether we experience these substances as nasty or nice depends upon how the taste cells are wired to our brain. Researchers used genetic engineering to switch the normal receptor for sugar on sweet-taste cells for a bitter receptor. Mice engineered in this way then reacted to bitter substances as though they were sweet and lapped them up rather than avoiding them as they normally would. It is this mechanism, with its variety of locks (receptors) for one gate (the taste cell), that makes it possible for evolution to tune the sensitivity of bitter taste to a wide range of molecules just by making slight changes to the receptors.

The sour ugly sister is not as unpleasant as her more complicated bitter sibling and has a more prominent role in cuisine. Sour is the taste of mild acids such as citric acid in lemons and unripe fruit or acetic acid in vinegar. The sourness of unripe fruit performs the obvious function for the plant that it deters animals from taking them until the seeds inside are ready to be sent out into the world. Vinegar is also a biological deterrent, but from a different source.

When fruit drops from the bough or milk flows from the breast, any that is not consumed will begin to ferment as yeasts, and bacteria make a feast of the leftovers. Fermentation is the process by which, in the absence of air, microbes consume sugars and produce waste products such as alcohol (in the case of yeast) or lactic acid (in the case of lactic acid bacteria). Alcohol and lactic acid are not just waste microbial products, but are also weapons of warfare that check the growth of other yeasts and bacteria, preventing them from competing for the microbe's food. We use fermentation for the same end to preserve food by pickling. As you will know if you have brewed beer or made wine at home, an air lock is essential to the success of the fermentation. If air gets into an alcoholic fermentation, this changes the environment to one in which aceto bacteria can flourish, and they turn alcohol into acetic acid (vinegar).

Acid molecules come in different shapes and sizes, but they all have the property in common that when in solution they add hydrogen ions to the chemical environment. Hydrogen ions ($H^+$) require no complicated lock-and-key receptors to trigger taste in the way that sweet, umami, or bitter molecules do, but instead stimulate the appropriate taste cells just by entering through channels in the cell membrane.

High concentrations of acid can damage cells and are probably for this reason sensed as unpleasantly sour, but a mild sour taste, especially when blended with another taste such as salty or sweet, adds a pleasant tang—for example, in gazpacho, a chilled soup from Andalucía in Spain that is made with wine vinegar, or in one of my favorites, hot and sour soup from Szechuan in China that contains a vinegar made from fermented rice. Fruit juices would be cloyingly sweet and lack their refreshing quality without the tartness of citric acid.

Curiously, children between the ages of five and nine react differently to sour-tasting things than do babies or adults. Charles Darwin saw this in his own children, whom he noticed loved rhubarb and fruit such as gooseberries that were just too sour for adult tastes. Candy manufacturers exploit the same phenomenon and produce very sour products aimed at this age group. One explanation that has been suggested for this preference is that it encourages children to eat fruits that contain vitamin C, though this hypothesis does not explain why the preference is lost in later childhood. Another hypothesis is that preference for sour-tasting food is not advantageous in itself, but is just an example of the desire to try new foods at an age when food habits are being formed for the future. This hypothesis is supported by a study that found that children who liked sour tastes the most were also less picky eaters and were more willing to try new food. Whether this would constitute an evolutionary advantage or not it is hard to say.

When we say that something is "a matter of taste"—whether it is a liking for salty anchovies, the color pink, or free-form jazz—we mean that people differ in what they like. It turns out that this is much more than a metaphor when it is applied to taste itself because the ability to taste is often influenced by genetic differences among people. Though there appears to be rather little genetic variation among people in the two umami receptor genes, the sequence of the *T1R2* gene varies in a way that suggests that it may be adapted to detect different sweet substances in different populations. However, the variation in all the *T1R* genes is low compared with that found in the genes that determine how different individuals taste bitter substances.

The best known example is taste for a chemical called phenylthiocarbamide, or PTC. To some people this chemical is intensely bitter, while to others it is nearly tasteless. This variation was discovered by accident in 1931, and it was quickly realized that being a "taster" or "non-taster" of PTC was something you inherited from your parents. Recent research has tracked down the genetic basis of this variation to differences in just one of the *TAS2R* genes: *TAS2R38*, which occurs in two alternate forms, or "alleles."

The interesting evolutionary question is why does this variation in *TAS2R38* exist? Two important facts about the PTC polymorphism suggest that evolution has conserved variation in *TAS2R38* for some reason. The first is simply that worldwide, 45 percent of people are non-tasters, which is an improbably high figure if tasters are at an advantage over non-tasters or, equally, the reverse. So could something be holding this variation in balance? This idea is supported by a curious discovery made by three of the founding fathers of modern evolutionary biology—Ronald Fisher, E. B. Ford, and Julian Huxley—while attending the International Congress of Genetics in Edinburgh in 1939.

While at the meeting, Fisher, Ford, and Huxley had the idea to visit the Edinburgh Zoo to find out whether the chimpanzees kept there were polymorphic (variable) for their reaction to tasting PTC. Remarkably, they found that they were. This can be interpreted in two ways. If the common ancestor of chimps and humans was polymorphic for the *TAS2R38* gene and both species have inherited the polymorphism from that ancestral population, this would mean that it has persisted for more than 6 million years. Alternatively, the polymorphism may have arisen separately in the two species, which would mean that there had been convergent evolution, perhaps with similar selection pressures at work in the two species. At the time of writing, the jury is still out on which of these is correct, but either way it is difficult to avoid the conclusion that natural selection really cares about variation in this particular gene for some reason. What could the reason be?

There is a clue in the differences in the genetic code between the two alleles. As we have seen in taste genes in other species, mutation can change the code and inactivate them if they no longer function to the advantage of the individual carrying them. Thus cats lost their ability to taste sugar, vampire bats are without umami taste, and all that is left of the relevant genes are ghosts of their once-working selves. However, this is not what has happened to the non-taster allele of *TAS2R38*. The mutational changes typical of a ghost are absent and the gene still functions—it just doesn't do what its taster version does. It still seems to make a bitter receptor, but not one that can be unlocked by PTC. Evolution has changed the locks.

There are 25 bitter-taste *TAS2R* genes in humans, and in most cases we still do not know which of the many bitter compounds match the receptors that they make, so it's not so surprising that in the case of the *TAS2R38* gene, we only have half the story. All 25 of the *TAS2R* genes are polymorphic, having multiple alleles, to some degree, but none shows a polymorphism like *TAS2R38* that is worldwide and so evenly balanced. Many plant compounds that taste bitter to us have medicinal value. Quinine is anti-malarial, for example. There is evidence that the bitter chemicals found in cucumber and its relatives such as zucchini—which have mostly been removed during domestication but can still appear in the fruit of some varieties when the plant is subject to drought stress—have anti-cancer properties. So it could be that the non-taster allele of *TAS2R38* confers on its carriers some important protective function by allowing them to eat more greens. Which greens, though, is the unanswered question, and what is the balancing advantage of the alternative taster allele?

I admit that the soup that I have served you in this chapter has contained more alphabetti spaghetti than liquor, but it is the suspension of the one in the other that makes taste. Taste sensations, like all biological processes, depend on a liquid medium. There is no such thing as solid-state biology. Evolution equipped our lineage very early on with taste receptors that tell nice from nasty and hardwired us to react accordingly. Comparing our taste receptors for umami, sweet, and bitter with those of other animals shows that we, like they, are equipped for our own particular diet. Fat is probably a basic taste too. It is certainly tasty. So our taste receptors signal our brains when we put essential nutrients into our mouths: protein (umami), carbohydrate (sweet), and lipid (fat). Taste receptors are, of course, just one part of the sensory equipment with which evolution has furnished us. Your sense of smell will tell you what is coming next to the table.

# 6

## Fish—Flavor

The flavor of fish can be a subtle thing or a roaring assault on the senses. Whether fish woos or wounds depends almost entirely on how fresh it is. The freshest fish is nearly odorless with a grassy aroma that is produced as polyunsaturated fatty acids are broken down by enzymes released by the fish's own cells. Fish flesh begins to decompose even at low (not freezing) temperatures that successfully preserve meat. This is because deep-sea fish tend to live at low temperatures anyway, so their enzymes are adapted to work in such conditions. A little more time for the enzymes to work and they release amino and nucleic acids, like our umami friends glutamine and inosine. A Japanese technique is to wrap fresh whitefish fillet in seaweed. In a couple of days in the fridge, the fish absorbs glutamate from the seaweed, which, coupled with inosine from the fish itself, enhances the umami taste of the fish that can then be enjoyed as sashimi.

Unless decomposition is checked by freezing, bacteria quickly join the feast and their activities produce increasingly more smelly molecules, leading flavor down a path from fresh to flat, sweet, then stale, and finally to putrid. According to Benjamin Franklin, "Fish and visitors smell in three days." The fishy smell is a compound called TMA (trimethylamine), which is the breakdown product of odorless TMAO (trimethylamine oxide). In its turn, TMA breaks down, releasing ammonia, another pungent component of fishiness. TMAO performs the same function in fish that sodium glutamate performs in seaweeds, achieving an osmotic balance with salty seawater that would otherwise suck the water out of their cells.

The five basic tastes—sweet, bitter, sour, salt, and umami—are scarcely adequate to completely capture any stage of the fickleness of fish flavor. This is because flavor is a multi-sensory experience that combines the five basic tastes with smell, touch (mouth feel), sight, sound, and memory to give us the infinite possibilities that we experience. Even pain receptors in the mouth contribute to flavor as it is through these that we sense the bite of the chili.

An eighteenth-century French chemist and priest named Père Polycarpe Poncelet was one of the first scientists to appreciate the complementarity between taste and smell. He drew an analogy between how different tastes can complement each other and musical harmonies, representing them on a musical stave. Smell is essential to flavor, and when this sense is disabled by a blocked nose caused either by a cold or just by pinching it shut, we enter an almost flavorless world that is flat and monochrome compared to the full flavorsomeness of everyday experience. But smell is the Cinderella of the human senses, under-appreciated by many and maligned since Aristotle, who more than 2,000 years ago wrote: "Our sense of smell is inferior to that of all other living creatures, and also inferior to all the other senses we possess."

It is certainly true that a bloodhound can follow a scent that to its human handler is totally undetectable, but is our sense of smell really inferior to that of all other living creatures? Allowing Aristotle some license for hyperbole in his attempt to make a point, can his claim be even approximately true? If odor is essential to flavor, creating so rich a vein of sensation, can our sense of smell really be so weak? Has evolution robbed us of the birthright that dogs and mice have inherited from our mammalian common ancestor that lived some 95 million years ago? What do the genes have to say about all this?

The sense of smell, like our sense of taste, is a chemical detection system and works in a similar way to the detection of bitter-tasting, sweet, and umami-tasting molecules. Smells, like all our other sensations, are perceived in the brain, which is wired by nerves to millions of olfactory (smell) receptor cells inside the nose. Like the receptors for bitter taste on the tongue, each olfactory receptor cell in the nose carries proteins on its outside called olfactory receptors (OR) that are triggered by only a limited range of molecules. Different receptor proteins are produced by different genes. Beyond that point there are some important differences between how taste and smell work.

While we have about 35 different receptors and as many genes for bitter-tasting substances, we have more than ten times that number of different olfactory receptors. We have about 400 different genes, each making a different OR protein. However, there is an even more important difference between bitter taste and olfactory receptors. Even though there are 35 different bitter receptors, we perceive the wide range of chemicals that trigger them as all tasting the same—bitter—because all the bitter receptor cells are wired into one line to the brain whose single message is "Urghhh." The olfactory receptor cells are not wired like this. Instead, each one of the 400 kinds has its own labeled line to the brain. This is like the difference between, say, having 35 phone lines that all go to the fire department where the message is always "Help! Fire!" and having phone calls from 400 friends, each of which carries a message that is as individual and different as they are. From an evolutionary point of view, it makes sense for the alarm system to be wired with a single line, but to have a more informative system for the perception of smells that carry much more subtle and varied information about food or sex.

So, was Aristotle quite wrong to claim that the human olfactory system is the poor relation of the senses and that we are the dullards of smell among our fellow creatures? The answer to this question is interesting and not as straightforward as at first it would appear. When we make a comparison with the number of OR genes that other mammals have, Aristotle seems to have been correct. African elephants, for example, have an incredible 2,000 functional OR genes, which must make them the smellingest animals on Earth. Try telling any friends who pride themselves on their discriminating wine-tasting abilities that they smell like an elephant and see how they react. It should be taken as a compliment.

When you consider that there are only about 25,000 genes in the entire human genome, and similar numbers in other mammals, having 2,000 OR genes or even half that number as rats and mice do suggests that evolution finds smell rather important, even in us humans with just 400. But why do we have so few OR genes compared to other mammals? Is it because the better-endowed species acquired more OR genes than us during evolution, or did we lose functional OR genes in the course of our own descent from the common ancestor of mammals? The answer is that there has been a great deal of evolutionary change in both directions. While we took the low road, elephants ascended the high path.

Not just humans, but other primates also have relatively few OR genes; chimps have similar numbers to ourselves, while orang-utans have fewer than 300. If the friend who prides himself on his olfactory abilities doesn't like the idea that he smells like an elephant, he might find comfort in the knowledge that at least he smells better than an orangutan. We can be sure that the low numbers of OR genes in primates represent big losses in our evolutionary history because primate genomes contain as many pseudogenes as intact working ones. In other words, our long-ago ancestors had many more OR genes than we do.

Pseudogenes are the remains of once-functional genes, like the rusted wrecks of old cars lining a highway, totally obsolete and going nowhere. It is certainly curious that while primates seem to have gotten along just fine with fewer and fewer OR genes as time progressed, African elephants belong to a group of mammals in which natural selection favored the multiplication of the very same class of genes. We don't know for the most part what our missing OR genes do in the animals that still have them, though we do know that they must confer powers of smell that we lack. Mice, for example, can smell carbon dioxide gas. So to a mouse, a sparkling mineral water must have a flavor that we simply cannot detect. Why there is this kind of difference between species is a scientific mystery, though it is probably related to differences between species in what they eat. We can hazard a guess as to what may have happened during the evolution of primates.

Natural selection weeds out the kinds of mutation that turn good working genes into broken wrecks, so long as those genes perform a function that directly or indirectly helps their carriers leave offspring. This suggests that if we look at how OR genes work, we might find a possible clue as to how they can lose their usefulness. Or, to put it another way, how many OR genes do we actually need? This is where it really gets interesting to the lover of food. The OR cells, which occur in a small patch of membrane in the nasal cavity, are exposed to smells that can come from either of two directions—from the outside via the nostrils, or from the interior of the mouth via a passage that connects the nasal passage with the back of the throat. The first direction is the one you use when you breathe in or sniff something and is called the orthonasal route. The second is called the retronasal route and is used in breathing out. It is the retronasal route that wafts all the volatile compounds that are released from your food when you chew it across the OR cells in your nose. This is what contributes the olfactory component to the creation of flavor.

Orthonasal and retronasal smells serve different functions. Orthonasal smells sample the outside world and tell you what is out there. Retronasal smells sample the intimate environment of the mouth and whatever you are eating or drinking. Even though retronasal smells are sensed in the nose, a mind trick causes us to experience them as flavor in the mouth. This is a major reason why we underestimate the power of our own sense of smell. We unconsciously attribute retronasal smells to taste and flavor rather than to olfaction.

It has been proposed that the primate loss of OR genes began when our ancestors started to walk on two legs and were able to rely on vision rather than smell to warn of danger. This would have decreased the importance of orthonasal smell, leaving the retronasal route as the more important one. The question then is are 400 OR genes enough to give us all the discriminating power that we need to tell the flavor of food that is nutritious from food that is harmful?

We can be certain that the answer to this question is yes because the brain does some very sophisticated processing of the inputs from the 400 types of olfactory receptor cell. The simplest thing a brain could do with 400 labeled lines of information would be to distinguish 400 flavors, but the brain is much, much cleverer than that. What actually happens is that many molecules trigger more than one kind of OR cell, and most OR cells respond to more than one kind of molecule. The result is that the brain never gets a ping on just one of the 400 labeled lines at a time, but invariably on a combination of several at once, and it is the specific combination that tells the brain which molecules are present in the nose.

Some OR cells respond to low concentrations and others to higher concentrations of the same molecule. Consequently, a different combination of OR cells may be stimulated and evoke very different reactions, depending upon on how much of the molecule there is. For example, the molecule skatole is found in the essential oils produced by jasmine flowers and orange blossom and also in mammalian feces. Low concentrations of skatole produced by flowers smell sweet and fragrant, while high concentrations of the same molecule emitted by feces smell foul.

How we perceive color provides an excellent example of what the brain can do when it combines even a few sensory inputs. In the retina of the eye, color is sensed by just three types of receptor cell, one tuned to respond to red light, one to blue, and one to green. By combining inputs from just these three cell types, the brain sees millions of colors including some like magenta that are pure inventions of the mind and totally absent from the rainbow spectrum.

So, Aristotle was both right and wrong about our ability to smell. He was right that we are not as well equipped as most other mammals to sniff danger or opportunity in our environment, but he was wrong on every other count. Thanks to our big brains, we are able to process the combined olfactory signals from a depleted repertoire of 400 receptor cell types into more than a trillion distinct smells. This makes olfaction hugely more sensitive than our color vision.

Add the many signals sent by OR cells to the brain to the signals that come from the receptors of the five (or six) basic tastes, plus sundry other sensory inputs from food, such as the feel and sound of crunching an apple or a crisp corn chip, combine them in all the ways imaginable, and the result is an unlimited variety of flavors. Far from being the poorest of our senses as Aristotle asserted, smell is actually the best of them. Ironically, like Aristotle, we remain unaware of our own superlative olfactory abilities because they are used mainly via the retronasal route and are hidden by the bodily illusion that flavor comes entirely from the mouth.

Evolution seems to like playing with OR genes. These genes and, we must assume, their concomitant smells and flavors not only vary in kind and number between different species, but there is also a great deal of variation among individuals within species. The sequencing of the human genome, achieved in the year 2000, was a scientific milestone in our growing knowledge of ourselves as a species. Two competing scientific teams raced to complete the first draft of the genome—a publicly funded team and one funded by venture capital that was led by Craig Venter. A couple of years afterward, Venter revealed that what his team had sequenced was not *the* human genome, but *a* human genome— namely, his own. We talk about *the* human genome as though there were just one, but each of us has our own copy and each is slightly different, especially when it comes to OR genes.

A comparison of approximately 400 OR genes in the genomes of 1,000 people found that on average there were 10 different versions (alleles) of each gene in this sample. Each person has two copies of each gene—one inherited from mom and one from dad—and it turned out that on average half a person's OR genes had two different alleles. This means that although we have only 400 OR genes each, we have half again as much genetic diversity in our personal OR repertoire, or 600 alleles each. Every one of these alleles is used, one per OR cell, to make its individual olfactory receptor protein, so all 600 strut their stuff in your nose.

Preference for different smells and flavors is an individual thing, greatly influenced by personal experience and food culture, but variation from person to person in OR genes has an influence too. The culinary herb cilantro (coriander)—which is widely used in Middle Eastern, Asian, and other cuisines—tastes unpleasantly soapy to some people. To John Gerard (c. 1545–1612), who wrote an early and very popular herbal, cilantro was "a very stink-ing herbe" with leaves "of venemous quality." A study of nearly 12,000 people who were asked whether they liked cilantro or not found that dislike of the herb was associated, albeit weakly, with a mutation in a specific OR gene.

In the absence of strong odors, the flavor of fish is heavily influ-enced by other properties, particularly the texture and oil content of the flesh, which is almost all muscle. Texture and oil content vary between fish species, depending upon how evolution has adapted their muscles to the needs of different lifestyles. Unsur-prisingly, the lifestyle of all fish is dominated by the properties of the medium in which they live—water.

If you watch a typical fish swimming, particularly from above, you will see that it propels itself through the water by undulating its body. This motion is produced by flexing the muscles alter-nately on one side of the body and then the other. With their streamlined body shape, fish require little energy to move at slow and steady speed through the water. This kind of cruising motion is effected by muscles containing the red pigment called myoglo-bin, similar to hemoglobin in red blood cells. Myoglobin stores the oxygen required by constant swimming, and the fuel it uses is stored in the form of oil. Herring, mackerel, and sardines are familiar examples of oily fish with dark flesh. The muscles of a plump herring can contain as much as 20 percent fat.

While it is almost effortless to cruise through water, the medium is very resistant to sudden acceleration. You can test this for yourself in a bath or pool. Moving the palm of your hand slowly through water is easy, compared to the effort required to move it suddenly and very quickly. A sudden movement creates a wall of water in front of your hand that presses back upon it. For a fish, rapid acceleration can be a matter of life-or-death in the jaws of a predator. For the predator, it is the difference between catching a meal or going hungry. So, to exert the force needed to accelerate rapidly, a fish needs a lot of muscle power that can be applied in an instant. This is provided by the white muscle found in abundance in big predatory fish such as cod and other whitefish. Cod muscle contains as little as 0.5 percent oil and no myoglobin. Tuna, which are large predators that migrate thousands of miles, have pink flesh made of muscle that is intermediate in its properties between white and red.

The structure of fish muscle is important to how it cooks and how it feels in the mouth. The muscles of land animals and those of fish perform in different ways. Land animals must use their muscles to support their bodies against gravity, for which end they are organized into tightly bound blocks that pull against bones that operate like levers. Bony fish achieve neutral buoyancy in seawater by means of a gas-filled swim bladder, so their muscles' only task is propulsion. The flaked nature of tenderly cooked fish that falls apart in the mouth to yield its subtle flavor comes from the way fish muscle is contoured in overlapping layers that are adapted for producing the sinuous motion required by swimming.

As decomposition proceeds, fish become more smelly, but not necessarily inedible. There is a Norwegian delicacy called rakfisk, made by salting fish and burying them for months to let them ferment. The odor has been described as like a selection of smelly cheeses left in a pile of used soccer gear for a week. Sauce made from fermented fish is an essential ingredient in the cuisines of Vietnam and Thailand, and a very similar sauce called garum was used routinely in ancient Roman cooking. The oldest surviving cookbook from Roman times—attributed, though probably erroneously, to a gourmet called Marcus Gavius Apicius who lived in the first century AD—contains 465 recipes of which more than three-quarters include the use of garum.

The manufacture and use of Roman garum two millennia ago has been reconstructed from documentary and archaeological evidence. The finest-grade garum was made from the trashiest ingredients—the blood and intestines of fresh mackerel. These were mixed 4 parts fish to 1 part salt in stone vats with stone caps placed on top to keep the ingredients submerged in the liquid that rapidly issued from them. The combination of salt and the exclusion of air suppress the growth of bacteria and fungi, so fermentation in these conditions is due to enzymes released from the cells of the fish itself. Using guts, in which food decomposition normally takes place anyway, presumably provided an especially rich supply of digesting enzymes. After months fermenting in the sun, the salty liquid was drained from the vat and bottled for use in cooking. Like modern fish sauce, garum would have been very rich in the umami ingredients inosinate and glutamate.

Roman authors seem to have had a love-hate relationship with garum because of the putrid smell issuing from its places of manufacture. Garum production was banned from some Roman towns, concentrating its production in a few coastal sites such as Almuñécar in Spain, where the stone vats used in the process during Roman times can still be seen. It has been claimed that fish sauce manufacture was the only large-scale factory industry in the ancient world. Amphorae used to transport garum have been found in ancient shipwrecks all over the Roman Empire, the fish sauce even reaching the outermost edge at Hadrian's Wall in northern Britain. Like the most popular modern-day cooking sauces, garum could make its producers wealthy, including one Aulus Umbricius Scaurus, a garum tycoon from ill-fated Pompeii in Italy, whose branded terra-cotta sauce bottles have been found a thousand kilometers away in southern France.

When fresh, some of the most flavorful products of the sea are not finned fish, but shellfish, which include mollusks such as mussels and clams and crustaceans like crabs and prawns. One of the reasons why they are so tasty is that the cells of shellfish do not resist the shriveling power of seawater with tasteless TMAO as finned fish do, but instead contain free amino acids such as glycine. These amino acids perform the same physiological function in shellfish that TMAO performs in finned fish, but they stimulate our umami receptors and thus taste delicious.

Taste receptors and soup demonstrate how evolution and cuisine take care of the bare essentials of life. Olfactory receptors and fish show how both are also capable of subtlety. Sniff the air and what do you detect? Is that roasting meat?

# 7

# *Meat—Carnivory*

Consuming meat shaped our evolution. In the family festivities of chapter 2, we discovered how our ancestors began to eat meat and became omnivores. More than 3.3 Ma, someone in our ancestor Lucy's Ethiopian homeland used stone tools to strip meat from bones. That someone was probably a member of Lucy's species, *Australopithecus afarensis*, the hominin thought to be the immediate ancestor of the human genus, *Homo*. She was evidently an omnivore like us, eating meat as well as plants.

Meat and fish are the richest sources of protein available to us and can supply all the essential amino acids that we need and that our tissues cannot manufacture for themselves. Meat also contributes other essential components of a balanced diet that are difficult to obtain in adequate quantities from plants alone. These include iron, zinc, vitamin $B_{12}$, and polyunsaturated fatty acids that are essential to the development of the brain and other tissues. It is of course possible to live healthily on a well-chosen vegetarian diet, but the challenge that a vegan diet totally free of animal products presents to human nutrition demonstrates that we are adapted to omnivory.

All of the animals that are farmed for meat—including the ancestors of those industrialized creatures now raised in crates, batteries, and pens—were hunted in the wild prior to domestication. It is time to hear their evolutionary stories and to learn how profoundly we shaped them. Stone tools are not the only evidence of how the journey from hunting to husbandry began. There is corroboration from an inside source: the tapeworm.

As adults, tapeworms live in the guts of animals. Life there is easy—food is delivered to the door, and all there is to do all day long is to hang around and make eggs. But, as for all parasites, the problem for those eggs is to find and infect a new host. Tapeworms do this by intercepting the food chain of their host, infecting the animals that the host eats. Human intestines are infected by three species of tapeworm, one caught from cattle (*Taenia saginata*) and two from pigs (*T. asiatica* and *T. solium*). We get these infections by eating meat that contains the larval stage of the tapeworms, which burrow into the muscles of cows and pigs. Only meat-eaters can catch tapeworms. Only tapeworm larvae eaten by a meat-eater can complete the life cycle of the parasite.

Since cattle and pigs are domesticated animals, it used to be thought that the tapeworm infections that we catch from them must have started with the advent of farming some 10,000 or 12,000 years ago. However, evolutionary analysis has discovered that our association with these parasites goes back millions rather than just thousands of years. *T. saginata* and *T. asiatica* share a common ancestor with a tapeworm species that infects and passes between lions and antelope in Africa. This suggests that an ancestor of these two human-infecting tapeworm species made the leap into our ancestors when they began eating the same prey as lions. What began as a lion-antelope tapeworm life cycle gave rise to a human-antelope life cycle about 2–2.5 million years ago, indicating that our ancestors must have been regular meat-eaters before that date.

Sometime after our ancestors acquired tapeworm, maybe as long ago as 1.7 Ma, the single species we carried split into the two species *T. saginata* and *T. asiatica*. We don't know how this split—or speciation event as it is called—occurred, but since the alternate hosts of the two tapeworms species are different, it probably happened because the separation of the tapeworm life cycle into one passing through cattle and another passing through pigs allowed tapeworms to adapt to the different requirements of infecting and surviving in these different hosts. Further evolutionary research on the speciation of our tapeworms could tell us something about the human diet at the time it happened. For example, if the tapeworm speciation event occurred in *Homo erectus*, did this ancestor broaden its diet at that time to include wild pigs as well as antelope, or did different populations of *H. erectus* prey on different animals and later cross-infect each other?

Our third tapeworm species, the pork tapeworm *T. solium*, shares a common ancestor with a tapeworm found in hyenas. *T. solium* must have been acquired in a similar fashion to the other tapeworms when our early or pre-human ancestors were eating the same prey as hyenas on the African savanna. Another intestinal parasite, *Trichinella spiralis*—acquired by eating pig meat—evolved into an infection of hominins by a similar route. Perhaps the first meat our ancestors tasted was killed by hyenas and lions and then stolen from them by hominins wielding stone weapons or even fire. However it began, our meat eating led to a closer and closer relationship with our animal prey, resulting ultimately in animal husbandry and the domestication of cattle and pigs. Our long evolutionary association with the three tapeworms that infect us suggests that we infected farmyard cattle and pigs with the parasites when we domesticated them, not the other way about.

We have two defenses against tapeworm infection: hygiene, to break the life cycle at the point where pigs or cattle come into contact with human feces that contain tapeworm eggs; and cooking, which destroys the infective stage in meat. If you enjoy your meat on the rare side, then you are relying upon good hygiene in the food chain and upon meat inspection in the abattoir to keep you safe from tapeworm and *Trichinella*.

The long association between *Taenia solium* and our species has left its genetic mark upon the parasite, which seems to have evolved some tolerance of cooking. Cells, including our own, contain heat-shock proteins that protect them against sudden rises in temperature. The genome of *T. solium* has an abnormally large number of the genes that code for heat-shock proteins, suggesting that it is unusually well-protected against heat shock, compared to tapeworm species that infect wild animals. If, as seems likely, hominins have been cooking meat for at least 1.5 million years, the evolution of an increased number of heat-shock proteins in tapeworms is to be expected as it would increase the chance that the infective stage in meat would survive cooking and be passed on, completing the parasite life cycle.

To the early evidence of carnivory found in stone tools and recorded by the tapeworm, we can add the later direct testimony of cave art. Though the very earliest images depict not animals, but the outlines of human hands, these were hands that wielded spears against game, grasped stone tools to strip flesh from bone, and built fires to roast the meat. In places as far apart as Indonesia, Australia, and Europe, a thousand anonymous hands stretch upward upon cave walls proclaiming proudly, like eager infants at their first kindergarten roll call, "I am here!" Forty thousand years later we hear them, not as a distant clamor but as the familiar clarion call of people reaching out to us in a common humanity.

The handprints were stenciled onto cave walls by blowing a spray of pigment from the mouth. Forty thousand years before the invention of the aerosol can, these ancestors airbrushed themselves into our family album with tags as personal as a graffiti artist's moniker. Five thousand years later, the first recognizable animal was painted on a cave wall on the tropical island of Sulawesi in Indonesia. It is a plump female babirusa, a peculiar animal that the zoologist who gave it its scientific name called *Babyrousa babyrussa*: so odd he named it twice.

The babirusa is a pig, endemic to Sulawesi, but like no other pig known. It has two pairs of tusks, one emerging from the lower jaw and another pair from the upper. The lower pair are just enlarged, curved canine teeth, but the upper pair grow from tooth sockets that have rotated upward so that the tusks emerging from them grow out through the middle of the animal's face and curve backward over it. The function of these strange tusks is unknown—they are too brittle to be of use in fighting or defense. Local legend has it that these tusks serve as hammock hooks by which the babirusa hangs from a tree branch when it wants a nap out of harm's way. Rudyard Kipling couldn't have come up with a better explanation if he'd included the babirusa in his *Just So Stories*. Babirusa are omnivorous, eating nuts and fruits, especially mangoes. Males can reach 200 pounds in weight. Just imagine if you caught one napping—what a roast babirusa that has lived on a diet of mangoes would taste like!

Twenty thousand years ago, hunters living in southern Europe created what is arguably some of the finest animal art of all time. The famous cave paintings at Lascaux and Chauvet in France and Altamira in Spain were made by Paleolithic people living on the edge of the ice sheets that then covered northern Europe. The landscape was treeless but teemed with animals grazing an open habitat that must have presented a scene like the tropical grasslands of the Serengeti today, but with the temperature dial in winter set firmly to –20°C or below. This vegetation type is called the mammoth steppe.

In the huge Chauvet cave in southern France, discovered as recently as 1994, there are images of animals from this period drawn with extraordinary skill. A pride of sixteen lions bounds in pursuit of seven bison; a mammoth and three rhinos stand by. In another part of the cave, three cave bears are drawn, skillfully using bulges in the rock surface to add lifelike relief to their anatomy. The remains of hundreds of this now-extinct species show that cave bears, far bigger than grizzlies, habitually hibernated there. On the walls there are also painted horses, bison, ibex, reindeer, red deer, musk ox, giant deer, and aurochs (wild cattle). These were artists who knew the anatomy of their subjects and had probably eaten many of them. The animal bones that they left behind in caves like Chauvet testify to a preference for reindeer meat and for the marrow of the long bones, which were smashed open to extract it. This was a world where meat was plentiful and could supply all the protein that the cave artists and their families needed.

The artistic hunters who stalked the animals of the mammoth steppe did not exist on a diet of meat alone, but also gathered and processed plant foods. Starch grains found on a cobblestone, recovered from excavations in a cave in southern Italy, show that it was used by people living there 32,000 years ago to grind the seeds of wild grasses into flour. They parched the grains before grinding them, a process still used today to improve the flavor and the keeping qualities of grains such as oats. In summer there would have been berries, hazelnuts, and roots, but in winter such pickings would be gone. Perhaps their reindeer steaks were accompanied by seeds and roots foraged from the caches that wild field mice make in their nests. This is how Inuit of Alaska and Siberia once supplemented their diet, being careful to distinguish between edible roots and those that mice can eat but which are toxic to people.

As the world warmed and the ice sheets that covered the north of Europe and North America retreated, the vegetation changed. Forests replaced the broad-leaved herbs, grasses, and low shrubs of the mammoth steppe, and many of the animals that grazed it retreated, following the plants that provided them with food. Reindeer and musk ox are now only to be found in the far north. In Europe, red deer and, more patchily, horses survived the environmental change, but other plant-eating species like woolly mammoth, woolly rhinoceros, cave bear, and giant deer all declined to extinction. With most of their prey disappearing, so too did the biggest carnivores, including the saber-toothed cat, the American lion, and a variety of gray wolf whose fossil skulls suggest that it was specialized for catching and devouring bison and other large prey.

Throughout the dwindling mammoth steppe, the largest and most specialized animals disappeared. The species that have survived resemble remnant populations, containing less genetic diversity than can be found in the DNA recovered from the bones of populations that perished. In some cases, animals may have been helped on their way to oblivion by human hunters finishing off what climate change had begun. Isotopic analysis of human bones, which yields a signature of the food that was eaten through life, indicates that the all-time favorite food at this time was woolly mammoth. These particular hunters lived 30,000 years ago, but archaeological evidence suggests that mammoth was an evergreen favorite on the menu all over the mammoth steppe right up until the species and their eponymous habitat vanished.

There is a famous Chinese proverb that says that every meal is enjoyed three times: in anticipation, in consumption, and in recollection. Meals of mammoth were enjoyed four times: in anticipation, in consumption, in recollection, and in habitation, because the large bones of mammoths were used to make dwellings too.

Can it be a coincidence that the last redoubt of this persecuted animal was on remote Wrangel Island off the northeast coast of Siberia, where they hung on until as recently as 4,000 years ago, more than 5,000 years after the last known continental population had disappeared?

Archaeological remains unearthed around the Mediterranean in Italy, Greece, Turkey, and Israel show that as long ago as 40,000–50,000 BP, people living along the coast began to broaden their diet, possibly because the human population was rising and, as a consequence even then, the largest game animals were becoming locally more difficult to find. Fast-forward 20,000 years and a detailed snapshot of life and diet before farming began is afforded by a remarkably well-preserved campsite called Ohalo II that has been found on the shore of the Sea of Galilee (Lake Kinneret) in Israel. Ohalo II was a temporary campsite that was regularly visited by hunter-gatherers over a number of years during the period that geologists describe as the Last Glacial Maximum, or LGM. This was the time when glaciers reached their farthest extent from the poles, and the climate of the Levant was cold and dry.

Twenty-three thousand years ago, Ohalo II became frozen in time when it was submerged by a rise in the level of the Sea of Galilee, covering it in mud and water. Waterlogged mud is a wonderful preservative because it excludes oxygen and this prevents bacteria and other agents of decay from destroying what lies beneath. This means that we not only have the bones that the Ohalo II campers threw away after meals and the buried skeleton of one of the male visitors, but also pieces of wood used in construction, the grass bedding people slept upon, cords and sinkers used to net fish, and the remains of the wild plants that were gathered and that are so rarely preserved at other archaeological sites of this age. More than 140 different species of plants used by people at Ohalo II or growing there have been identified. These include wild wheat and barley that we know were used to make flour because there is also an abandoned grinding slab with cereal starch grains preserved in its surface. The plant remains also include seeds of thirteen species that are today weeds of cultivation, suggesting that cereals may possibly have been deliberately sown.

Besides fish, which must have been the main attraction of the lake, people at Ohalo II ate large quantities of gazelle; every kind of bird they could catch, including grebes, ducks and geese, birds of prey, and crows, as well as deer and the very occasional aurochs, wild pigs, and goats. Remarkably, apart from aurochs that are now extinct and mountain gazelles that have become rare, most of the animals and plants that the people at Ohalo II ate or used are still to be found in the area today.

Ohalo II was abandoned when it became flooded, but we know from a later Israeli site near Haifa called el-Wad, that over the following 8,000 years hunter-gatherers continued to enjoy a similar diet with a broad range of prey. Mountain gazelle remained the favorite meat, but people at el-Wad also ate from an even broader wild menagerie that now included small animals such as turtles and even snakes. This increase in small animals in the diet probably signified that the local supply of larger game animals was beginning to suffer from overkill. The growing impact of hunting would certainly explain why, less than 4,000 years later, around 11,700 BP, remains of all animals, large and small, are scarce in the archaeological record at el-Wad. These kinds of changes occurred generally throughout southwest Asia, where people were beginning to rely more and more upon other means of subsistence.

The gradual domestication of plants like wild wheat, barley, and pulses that had been proceeding for millennia (chapter 4) could provide protein-rich crops to substitute for the failing supply of wild meat, but there was another solution as well. We can discover what people ate from the archaeological record preserved inside artificial mounds that were created by millennia of settlement as mud houses were built, decayed, and the remains were built upon. Each new layer of mud houses entombed a record of previous habitation and eating habits in its foundations. Excavations in such a mound at Aşıklı Höyük in Anatolia, show that between 11,000 BP and 10,200 BP, hunters switched from a wild-meat diet like the one once eaten at el-Wad to rearing sheep. This shift to a reliance on domesticated animals and plants marked the beginning of agriculture and the dawn of the Neolithic.

The change from hunting and gathering to farming as a means of subsistence triggered a Neolithic baby boom. It is estimated from human skeletal remains in Levantine cemeteries of the period that the number of children born per woman nearly doubled from 5.4 among hunter-gatherers to 9.7 when farming was adopted. Such increases were a global phenomenon that occurred wherever hunter-gatherers switched to farming. Because farmers hunt and convert natural habitat to farmland, the effect of the Neolithic population increase was to increase still further the pressure on wild animal populations and to favor animal husbandry as a means of feeding all the extra mouths. The practice of domestication began a new chapter in our evolutionary impact upon nature and altered the course of our own evolution too.

Some animals made the transition from hunted prey to domesticated farm animal more easily than others. In evidence of this, a few species like the wild pig and the chicken were domesticated several times, while many others, like the dozens of species hunted by the inhabitants of Ohalo II, were never tamed. Pigs and chickens are scavengers, which habit perhaps brought them into close proximity with human habitation, where they became reliant on people for food, initiating the process of domestication.

Chickens are the most portable of our domestic livestock, and we have carried these amazing creatures, which can turn scavenged scraps into tasty meat and a daily supply of eggs, everywhere. If there were an avian Homer, he could apply his quill pen to writing an odyssey of the travels of the chicken, but until he shows up we'll have to make do with Dr. Seuss, who presciently warned young travelers what to expect: "You'll get mixed up, of course, as you already know. You'll get mixed up with many strange birds as you go."

Movement around the planet has evolutionary consequences, both for the organisms making the journey and for the residents with whom the new arrivals get mixed up. The domestic fowl that looks so at home in farmyards and backyards, whether in New England or old England, has a much more exotic origin. Exactly as Charles Darwin conjectured with customary insight in his book *The Variation of Animals and Plants under Domestication*, the wild ancestor of the chicken has proved to be the red junglefowl, native to Asia.

Modern genetic and archaeological analysis shows that the red junglefowl was domesticated not just once, but three times independently in different parts of Asia (see map 5). The earliest archaeological evidence of chicken use is from the valley of the Yellow River in northern China. Ancient chicken bones found there date from 10,000 years ago, and DNA extracted from them shows a genetic affinity with modern domestic chickens. There is no direct evidence that these early chicken bones belonged to domesticated rather than to wild birds, but other animals such as the pig were being farmed in the area around the same time, so it seems very likely that domestication would have taken place then or soon afterward. Somewhat after the Yellow River domestication, red junglefowl were independently domesticated around Thailand in Southeast Asia and also in India.

The genes of modern chickens reveal a complicated ancestry that is testament to the traffic around the globe that has brought the descendants of all three domestication events into fertile contact with one another on many occasions. The first recorded pan-Asian meeting between chickens occurred thousands of years ago when Buddhist monks from China brought a live chicken home from India as an edible souvenir.

The yellow skin preferred by consumers in the United States and elsewhere is not found in wild red junglefowl at all and appears to have come from another species—the grey junglefowl. This species is native to India but does not hybridize with the red species in the wild, so the domestic chicken probably acquired the yellow skin trait from its grey cousin in a farmyard somewhere in India many thousands of years ago.

The chickens of Africa—which according to the Food and Agriculture Organization of the United Nations numbered 1.6 billion birds in 2010—arise from at least three separate introductions to the continent. One journey brought Indian stock by a northern route from Egypt, where chickens are mentioned in ancient texts that are 4,000 years old. A second seems to have arrived from the east via the Horn of Africa, following a reverse route to the one

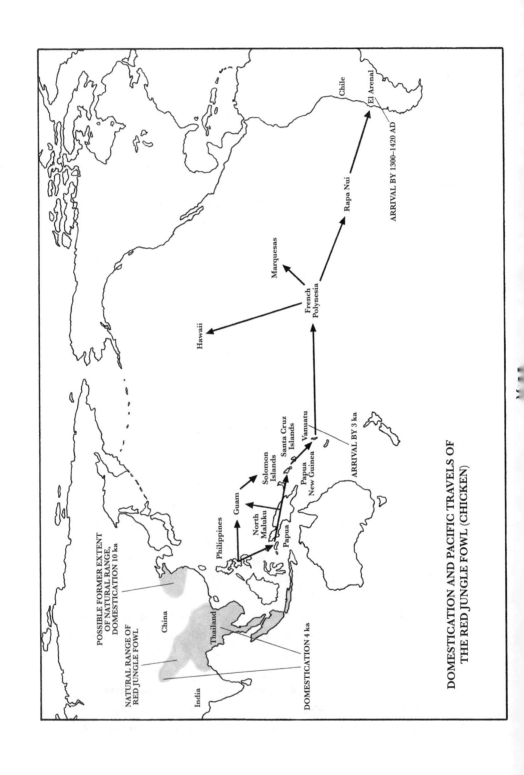

POSSIBLE FORMER EXTENT
OF NATURAL RANGE,
DOMESTICATION 10 ka

NATURAL RANGE OF
RED JUNGLE FOWL

China

India

Thailand

DOMESTICATION 4 ka

Philippines

North
Maluku

Papua

Guam

Solomon
Islands

Santa Cruz
Islands

Vanuatu

Papua
New Guinea

ARRIVAL BY 3 ka

Hawaii

Marquesas

French
Polynesia

Rapa Nui

Chile

El Arenal

ARRIVAL BY 1300–1420 AD

**DOMESTICATION AND PACIFIC TRAVELS OF
THE RED JUNGLE FOWL (CHICKEN)**

that humans took out of Africa. The third also arrived from the east but came by sea from Southeast Asia.

Southeast Asia was also the point of departure for the Polynesians who peopled the Pacific islands, taking live chickens, rats, dogs, and food plants with them as they went. This was perhaps the most heroic of all migrations in human history, covering thousands of kilometers of open sea by canoe, reaching the remotest islands of the Pacific—Rapa Nui (Easter Island), Hawai'i, and ultimately New Zealand—where they arrived comparatively recently, only 800 years ago.

Rapa Nui is famous for its giant statues that stand sentinel over a barren, treeless environment that seems scarcely capable of supporting a population with the numbers, the leisure, or the inclination to build such monuments. However, other structures on the island give the lie to this impression—the 1,233 stone chicken houses that rim the coastline. The chicken houses range in length from 20-foot, household-sized ones to 70-footers, 10 feet wide and 6 feet high, with a little entrance at ground level for the chickens to run in and out of. A stone enclosure around each house protects a chicken run. The chicken houses of Rapa Nui indicate that the birds, which were the only livestock available, were farmed on an industrial scale and must have provided a ready supply of meat and eggs.

The food that the Polynesians transported with them in their oceangoing canoes provides evidence that they reached the Americas before Christopher Columbus. Ancient chicken bones dated to between 1300 and 1420 AD have been excavated at El Arenal in south-central Chile. Analysis of the DNA in these bones has shown them to be genetically almost identical to prehistoric Polynesian chickens that have been excavated on Samoa, Tonga, and Rapa Nui. Thus, the chickens that the Spanish conquistador Francisco Pizarro saw when he arrived in Peru in 1532 were an established part of the Inca economy and probably originated from Polynesia.

Despite the thousands of kilometers traveled, the arrival of the chicken in the Americas as a passenger in a Polynesian canoe was not the chance occurrence that it may sound. Polynesian seafarers knew very well where they were going. They were extraordinary navigators who used a detailed knowledge of the stars of the night sky for navigation, together with the swell of the ocean as it was deflected by land to detect the direction of islands that were far out of sight. The most advanced practitioners of this art entered the water and used the motion of the ocean upon the scrotum as a direction finder. So, Polynesian canoes ran on ball bearings a thousand years before wheels did.

One plant species was transported back from the Americas to Polynesia, indicating that the ocean traffic was two-way. When Captain James Cook first explored the Pacific in 1769, he found that the sweet potato, originating from South America, was being grown on all the Polynesian islands that he visited, including remote New Zealand. Recent genetic analysis of the DNA from sweet potato specimens collected by botanists on Cook's voyage has confirmed that the plants they found in Polynesia originally came from the region of Ecuador and Peru in South America.

As might be expected if there was significant traffic between South America and Polynesia, there is evidence of contact in the human genome too. Since they were "discovered" by Europeans, the Polynesian natives of Rapa Nui have been decimated by disease, invaded, enslaved, and transported, but the genomes of the surviving remnant population contain a historic signature of happier times when visitors from South America were made welcome and became part of the family.

Returning to our quest for the roots of animal domestication, among mammals social behavior seems to have predisposed some species to this end. Social animals like sheep that live in herds or dogs that live in packs in the wild were readily domesticated, while harem-forming animals such as deer were not. The very earliest domesticate was the dog, which evolved from the gray wolf and became our companion in the hunting of other animals at least 15,000 years ago, although it has been suggested that the dog-human association could be at least twice as old as this. Dogs seem to follow and obey their human handlers as they would the leader of a wolf pack. You only have to watch a shepherd commanding a sheepdog as it rounds up sheep to understand the importance of the social behavior of both animals to sheep farming. Managing herds of sheep with dogs is just another example of how we exploit existing evolutionary relationships for our own ends.

Sheep were early domesticates in southwest Asia, maybe as early as 11,000 years ago, and spread from there in several waves to all points of the compass. By 5,700 years ago, sheep of southwest Asian origin had reached as far away as northern China. The world sheep population is now more than a billion animals. Wherever sheep wound up, they were bred and became adapted to local conditions, so that now there are about 1,500 different breeds. Sheep evolution did not stand still back home in southwest Asia, either. Sheep are capable of accumulating large amounts of fat, and some thousands of years after domestication and the first wave of dispersal from southwest Asia, farmers there produced breeds of sheep with a spectacularly fat tail that began a fresh wave of dispersal. The Greek historian Herodotus (c. 484–425 BCE) wrote that the tails of some Arabian sheep were so large that shepherds harnessed small wooden carts to their animals so that they could drag their tails around without damaging them. The fat from the tails of these sheep is a traditional cooking medium used in the Middle East and Iran. Sheep whose tails have been docked can partly regrow them.

Wild pig, or boar (*Sus scrofa*), and wild cattle, or aurochs (*Bos primigenius*), were until recently both distributed right across Eurasia from the far west to the far east, affording many societies in this vast geographic range the opportunity to domesticate them. In southwest Asia, the transition that took place between hunting wild cattle and pigs and rearing them has been detected at a Neolithic settlement in the upper Jordan valley, where, during the interval from 9000 to 8000 BP, bone deposits typical of domesticated animals become more common as those of wild animals of the same species become rarer.

By 8000 BP, both cattle and pigs were fully domesticated in southwest Asia, but the genes of the two species tell very different stories about what happened thereafter. In the case of cattle, genetic analysis of living animals tells us that the aurochs was domesticated three times: in southwest Asia, quite probably in Syria; in the Indus valley, where the aurochs gave rise to the domesticated zebu with its characteristic hump; and in Africa.

The aurochs was a common wild animal in Europe as recently as Roman times, but European cattle breeds all descend from ancestors that were domesticated in southwest Asia, rather than from local European populations of aurochs. Studies of human genetics show that farming itself spread into western Europe with a migration of farmers from southwest Asia, so it appears that they took their cattle with them. Farmers, farming, and their cattle dispersed into Europe as a package.

In complete contrast, the spread of cattle and other domesticated livestock through the Fertile Crescent itself (see map 6) was not accompanied by human migration. The animals spread from farming community to farming community, but genetics tells us that the people stayed put. So much so, in fact, that genome sequences recovered from 9,000-year-old remains in the Zagros Mountains of Iran show that the Neolithic farmers of that time left a genetic legacy among Zoroastrians who live in Iran today. This stay-at-home habit was present throughout the Fertile Crescent. The first farmers in Anatolia, the Levant (Israel and Jordan), and the Zagros Mountains traded with each other and shared farming practices, livestock, and crops, but they did not share blood ties.

In Africa, wild cattle were the first animals to be domesticated. They were then interbred with cattle from southwest Asia and India to produce animals that are adapted to local conditions. Unlike the situation in southwest Asia where domesticated cattle were integrated into settled agriculture after the domestication of cereals, in sub-Saharan Africa, cattle formed the basis of nomadic pastoralism many thousands of years before the first indigenous crops were domesticated there. Cattle remain of supreme economic and social importance in many African cultures where a man's worth is counted in the head of cattle he owns.

The original evolutionary home of the ancestors of the wild boar, *Sus scrofa*, is on the islands of Southeast Asia, where the babirusa and other wild pigs can still be found. From the whole porcine family, *Sus scrofa* was the little pig that went to market and spread westward into Eurasia, arriving millions of years before our own species reached there from Africa. Later, wherever humans and pigs encountered each other, they formed a bond.

# THE DOMESTICATION AND SPREAD OF FARM ANIMALS IN THE FERTILE CRESCENT

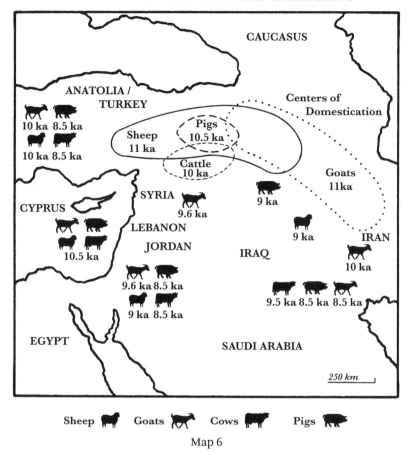

Sheep 🐑  Goats 🐐  Cows 🐄  Pigs 🐖

Map 6

Pigs and people can share an affinity almost as close, though admittedly not as universal, as that between people and dogs. Dogs and pigs are both scavengers, which is probably the foundation of their relationships with our species and the route by which each of them became domesticated. This affinity shaped the evolutionary history of the pig, which entered domestication not once like sheep or even three times like cattle but six or seven times at least.

In western Europe, in contrast to the domesticated cattle of this area that came from southwest Asian stock, pigs were domesticated from indigenous European wild boar. Likewise, pigs were independently domesticated from the local wild boar on the Mediterranean islands of Sardinia and Corsica, at least twice in China, and also in Burma and Malaysia. The wild boar of New Guinea are probably feral animals descended from domesticated pigs transported there in canoes by the Polynesians. The source of the Polynesians' pigs, which they carried with them to some of the remotest islands of the Pacific such as Hawai'i, appears to have originally been Vietnam.

Curiously, the one place where the genes of local wild boar do not show up in modern domesticated pigs is in southwest Asia, the very source of so many other domesticated animals and plants. This is doubly strange because we know from the archaeological record that pigs were hunted and then domesticated in southwest Asia in Neolithic times, but for some reason the genes of modern domesticated pigs show no affinity at all with this area. The reason may be an historical and cultural one. The majority inhabitants of southwest Asia today are Muslims and Jews, for both of whom the pig is an unclean animal forbidden by dietary law. This religious taboo may have been passed down from ancient Egypt, where there was an on-off, love-hate relationship with pigs. They began as venerated animals, but by 1000 BCE they had become associated with the god Seth, the evil pig-faced foe of the sun god Horus, who was blinded by a black pig. The image of Seth was expunged from monuments, and swineherds were so despised that they were not allowed to enter any temple. In this cultural and religious milieu, it seems less surprising that local breeds of pig might go extinct.

While their social behavior predisposed compliant animals like pigs, sheep, and cattle to domestication, it made others recalcitrant to it. Territorial animals like deer and antelope have never been domesticated, which is why the mountain gazelle never occupied the Neolithic farmyard, even though it was a favorite of the hunt. Likewise, red deer have been hunted in Europe for 50,000 years, but have never been conventionally farmed because they are territorial and the rut, in which males contest for mates by battling each other, makes them difficult to manage. Reindeer are the exception that proves the rule in that they are the only non-territorial deer and have been domesticated twice, by the Saami people in Lapland and by the Nenets of Russian Siberia. Both the

Saami and the Nenets are nomadic, following their herds of reindeer as these roam the tundra in search of food. The relationship is almost a commensal one, like that between humans and dogs, but in reverse since humans follow the reindeer herd and dogs follow humans.

In *The Variation of Animals and Plants under Domestication*, Darwin conducted an encyclopedic survey of how domesticated livestock differed from their putative wild ancestors, revealing a remarkable pattern that neither he nor anyone else until very recently has been able to convincingly explain. Darwin observed that all kinds of quite unrelated domesticated animals— including dogs, pigs, cattle, rabbits, guinea pigs, and horses— shared a tendency toward a common set of traits. Darwin noticed that compared to wild animals, domesticated ones were less seasonal in their pattern of breeding, often had piebald coats that lack pigmentation over parts of the body, possessed floppy ears, shorter muzzles, smaller teeth, smaller brains, curly tails, and were more juvenile and docile in their behavior. This set of convergent traits is now known as the domestication syndrome, and it has taken nearly a century and a half for anyone to offer a plausible explanation as to why domestication should create such a repeatable collection of seemingly miscellaneous evolutionary changes.

Evolutionary change is often due to selection, and when it results in convergence between unrelated animals, the simplest explanation is that the causes of selection must be the same in the different cases. This explanation is plausible for some traits in the domestication syndrome, but not for all of them. Docility is obviously a desirable trait in any domesticated animal, and most breeders wittingly or unwittingly select in favor of this characteristic, so it is not at all surprising that it is part of the domestication syndrome. In his novel *The Restaurant at the End of the Universe*, Douglas Adams conjures a scene in said restaurant in which the waiter asks, "Would you like to meet the dish of the day?"

A large dairy animal approached Zaphod Beeblebrox's table, a large fat meaty quadruped of the bovine type with large watery eyes, small horns and what might almost have been an ingratiating smile on its lips. "Good evening," it lowed and sat back heavily on its haunches, "I am the main Dish of the Day. May I interest you in the parts of my body?"

Arthur Dent, an Earthling far from home, recoils in horror and orders a green salad, much to the disgust of the Dish of the Day. "Are you going to tell me that I shouldn't have green salad," says Arthur, to which the animal replies:

" 'I know many vegetables that are very clear on that point. Which is why it was eventually decided to cut through the whole tangled problem and breed an animal that actually wanted to be eaten and was capable of saying so clearly and distinctly. And here I am.' It managed a very slight bow."

This is fiction, but only just. Domesticated animals are bred to be biddable, even if they are as yet unable to tell us so. But why do they also have floppy ears and curly tails? Why does the domestication syndrome include piebald pigmentation, found so prominently in such distantly related animals as cows, dogs, guinea pigs, and even koi carp? Since direct selection acting in parallel on such a disparate collection of animals and traits seems unlikely, there must be an alternative explanation. Could there be some common underlying genetic cause that links all these traits together, so that artificial selection on a trait such as docility somehow influences coat coloration along with all the other traits of the domestication syndrome? Far-fetched though this may sound, there is experimental evidence that suggests that this might be correct.

In the 1950s a Russian animal breeder named Dmitry Belyaev began an experiment to see whether the complete domestication syndrome would appear in Siberian silver foxes—which until that time had not been tamed—if animals were selectively bred purely on the basis of their docility. At the start of the experiment, nearly all the foxes showed aggression or fear when experimenters approached to feed them. The small fraction of animals showing the least fear or aggression toward their handlers were used to breed a new generation, and this process was then repeated over many decades. After only three generations, no fox cubs showed aggressive behavior when fed, and some even began wagging their tails like domestic dogs. After eight to ten generations of selection for docility, fox cubs with patterned coats, floppy ears, and curly tails began to appear. After five decades and more than thirty generations, the entire experimental group of silver foxes behaved with dog-like friendliness to humans and also displayed the anatomical and physiological traits of the domestication syndrome. This experiment demonstrated that the whole domestication syndrome does arise simply from selection for docile behavior, but how?

The Russian scientists who now continue the research begun by Belyaev have suggested that all the traits in the domestication syndrome must be controlled by a single network of genetic switches. This would be like some manic chariot driver with reins on a dozen wild horses, managing to steer all of the domestication traits in a coordinated fashion and getting them to go in the same direction. If there is such a control mechanism, then no one has yet been able to find it. Instead, three scientists—including Richard Wrangham, originator of the cooking hypothesis (chapter 2)—have recently come up with another explanation.

The hypothesis is that all the traits of the domestication syndrome are not linked by a genetic master switch, but by a common step in embryonic development. During the development of the vertebrate embryo, all the traits of the domestication syndrome depend, directly or indirectly, on the supply of cells from a single source called the neural crest. The neural crest runs from head to tail along the spine of the developing embryo and contains stem cells that supply the raw material for the construction, or the control of the construction, of the brain, the cells that make skin pigments, the adrenal glands that influence aggressive behavior, and other cells and organs relevant to the domestication syndrome.

The neural crest hypothesis proposes that due to the fact that the adrenal glands control aggressiveness, the less aggressive animals selected during domestication belong to the fraction of the population with genes for smaller adrenals. The genes controlling the size of the adrenals, and hence aggressiveness, do so only indirectly through their influence on the number of cells in the neural crest. Thus, according to this hypothesis, when more docile animals are selected, what is actually happening is that breeders are selecting for animals with an inherited deficiency in the number of cells in the neural crest. It is dependency on cells from the neural crest that links all the traits of the domestication syndrome together and causes them to evolve as a group of traits.

The number of neural crest cells is believed to be controlled by many genes, each with a small effect, which is presumably why no single master gene controlling the domestication syndrome has been found: there just isn't one. If the neural crest hypothesis is correct, then it would mean that most of the changes observed during domestication—such as curly tails, piebald coats, and floppy ears—are incidental by-products of the main event, which is selection for docile behavior. Though it is too early to say whether the idea is correct or not, there has been no other hypothesis since Darwin first wrote about domestication in 1868 that has so neatly explained why the syndrome exists.

Who knew that there was so much evolution to relish in the meat we eat? Let me recap and conclude. Our meat-eating is older than the human species, or even than the genus *Homo*. We have the testament of worms for that, and the fossils concur. But the plant remains at Ohalo II demonstrate very clearly that ancient hunter-gatherers living during the Last Glacial Maximum did not live exclusively on meat. Neither, for that matter, do the hunter-gatherers who live this way today. Nonetheless, it seems that it may have been the exhaustion of the wild meat supply as the human population grew, perhaps aided by the impact of a changing climate, that impelled people in southwest Asia, China, and elsewhere to begin cultivating crops and domesticating animals.

Today, the farmyard is a zoo in which are gathered together domesticated animals from all over the planet. We moved, milked, and molded them in ways so fundamental that a domestication syndrome appeared, marking animals as different as the pig and the dog with a uniform of piebald coats and floppy ears that brand them all as remade by humans. Despite the uniformity of the domestication syndrome, add the huge diversity of plants that is grown in farms and gardens, and there can be no doubt that variety is the watchword of the human diet. There is no better demonstration of this than in the many vegetables that we eat.

# 8

## *Vegetables—Variety*

Our pre-human ancestors were vegetarians, but we now consume a greater variety of plants than ever they could. More than 4,000 different species of plants are eaten as foods or flavorings, even though most of them have evolved poisonous chemical defenses against plant-eaters. The proverb that "one man's meat is another man's poison" would be more accurate if it read "one man's vegetable is another man's poison." Fresh meat is seldom poisonous to anyone, but vegetables, at least in the wild, almost invariably are. While we lack the capacious digestive tract of a gorilla or a chimp, two technologies make the extraordinary variety of plants in our diet edible: cooking and plant domestication.

Cooking not only renders tough foods tender, but can also make the toxic tolerable. For example, kidney beans contain poisonous lectins that in nature protect beans from the insects and fungi that attack them. Boiling kidney beans destroys the lectins; however, cooking at too low a temperature, in a Crock-Pot or slow cooker that does not reach boiling point, can soften the beans without inactivating the lectins with poisonous results. Domestication of the common bean has generated a gallimaufry of varieties such as white navy, mottled pinto, black and green flageolet, some of which no longer contain poisonous amounts of lectin.

Wild plants have characteristics adapted to their own environments and needs, which we subvert for our own ends through domestication. For example, wild potatoes make small tubers the size of a plum or a pea that are distributed around the plant on stolons a meter or more long. Clearly, in wild potatoes natural selection favors plants that spread, and so a plant's energy goes into making long stolons rather than big tubers. Through artificial selection we have reversed this to suit our own purposes, so that the varieties that we grow produce big tubers on short stolons right underneath the plant where we can easily dig them up.

Like natural selection that produced life in all its amazing richness, artificial selection can work wonders with just the raw material of heritable variation. Consider what plant breeders have managed to create from wild cabbage, a weedy and inedible-looking plant found around the coasts of northern Europe. From this unpromising beginning, centuries of selective breeding by unknown horticulturalists who knew nothing about genetics or evolution generated cauliflowers ("nothing but a cabbage with a college education," according to Mark Twain), broccoli, Brussels sprouts, kohlrabi, and kale, not to mention cabbages with giant, compact heads. There is even a novelty cabbage bred in the Channel Islands, near the coast of Brittany, France, that produces a stem that is tall and strong enough to make a good walking stick and that used to be grown for just that purpose.

Natural and artificial selection are both gradual in their effects, but the latter can achieve changes much more quickly. The wild progenitor of the cultivated tomato has small berries that are dispersed by birds, but the biggest cultivated varieties such as beefsteak are a hundred times bigger. A great leap forward in tomato size and quality was made by an amateur grower in Baltimore called Dr. Hand, who in about 1850 began a program of crosses and selection that in little more than 20 years produced a huge, fleshy tomato of excellent flavor called Trophy. Thanks to genetics, which was a closed book to breeders in the nineteenth century, we now know, in principle at least, how Trophy and other traditional varieties were made.

The genetic variation that provided the raw material for tomato domestication and improvement existed in wild populations long before it was put to use by humans. The original domestication event and subsequent transport from place to place squeezed tomatoes through at least three genetic bottlenecks as each time only a small sample of tomatoes made it through. The first domesticated tomatoes in Mexico contained only a fraction of the genetic variation present in wild populations. Then, when tomatoes were taken from Mexico to Europe in the sixteenth century, only a fraction of the fraction made the voyage, and when European varieties came back to the Americas, the dilution was repeated. These bottlenecks reduced the genetic variation present in cultivation to less than 5 percent of that existing in the wild species, but, even so, this small fraction contained enough variation to provide the raw material for remarkable changes made by artificial selection.

Analysis of the tomato genome suggests that only a handful of genes were responsible for what Dr. Hand accomplished. His results were achieved by rearranging existing genetic variation among plants, or as a contemporary put it, Dr. Hand managed to put the bulk of a large misshapen tomato into the smooth, round skin from a small one "and then, by careful selection, year after year increased its size and the solidity of its contents." Domestication has often wrought big changes in crops through selection acting on single genes that regulate what other genes do. A regulatory gene is like the conductor of an orchestra, determining the pace and timing of the actions of many players. It is easier for artificial selection to influence the whole orchestra of genes through the conductor than by individually tweaking each player.

The Trophy tomato was in such great demand when it appeared in 1870 that the seeds fetched $5 for a packet of 20, which is the equivalent today of $100 a packet, or $5 for a single seed. For a short while, the nurseryman selling the seeds, one Colonel Waring, made a killing. He offered a $5 prize for every Trophy tomato sent to him that weighed 2½ pounds or more and $100 dollars to the person sending him the biggest and best Trophy tomato of all. He then purchased the whole of the winner's crop to resell as seed. With such astute marketing and, in effect, crowdsourcing the best seed from prize-winning growers, Trophy spread like wildfire. However, evolution rarely stands still, particularly in the hands of enthusiastic growers, and Trophy was selected and crossed to such a degree that within 20 years seed suppliers were complaining that they could no longer find true seeds of the original variety. Trophy's genetic legacy is in the hundreds of new varieties it spawned.

The diversity of heirloom varieties, produced before modern commercial breeding began, is mainly due to local adaptation and the idiosyncratic preferences of tomato growers like Dr. Hand. There are about a dozen species of wild tomato, but the progenitor of the cultivated tomato, *Solanum lycopersicon*, is the only one to have been domesticated, and this appears to have happened only once. The wild species is native to the Andes, but seems to have been ignored by the indigenous people who lived there, even though they were formidable plant domesticators. Instead, the wild tomato found its way north, possibly as a weed, and was domesticated by the Maya in Mexico. The wild tomato with cherry-sized berries grows as a wanted weed in Mexico to this day. It is not deliberately sown, but when it pops up in a field of its own accord, it is protected by farmers for its flavorsome Lilliputian fruits. This practice may have initiated domestication. The date of tomato domestication is not known, but by the sixteenth century, the Spanish priest Bernardino de Sahagún saw a huge variety of *tomatl* (the Nahuatl word from which we get "tomato") in the Aztec market at Tenochtitlán (Mexico City): ". . . large tomatoes, small tomatoes, leaf tomatoes, sweet tomatoes, large serpent tomatoes, nipple-shaped tomatoes," and tomatoes of all colors from the brightest red to the deepest yellow. The Aztecs taunted the Spanish invaders that they would end up in a dish of tomato and chili.

Geography adds variety to vegetables. New landraces evolve as growers select what suits their location and what takes their fancy. The individuality of these varieties is often evident in the names that evoke the personalities and predilections of the grower and the places where they were first grown. Aunt Ginny's Purple is a tomato with large pink fruit, which came from Germany and, according to the website offering it for sale, was successfully grown by a family in Indianapolis for over 25 years. Aunts Gertie and Ruby also have their tomatoes listed on the same heirloom tomato website, along with Ethel Watkin's Best, John Lossaso's Low Acid Ruby, Livingston's Gold Ball, Middle Tennessee Low Acid, and the Missouri Pink Love Apple. Every common fruit and vegetable has a similar anthology of poetically named varieties that celebrate the confluence of human ingenuity, plant diversity, and locale.

The Andes, from where the wild tomato came, and Mexico, where it was domesticated, are both important regions where many new crops evolved. The Andes are the second-highest mountain range in the world, averaging more than 3,000 meters in altitude. In Peru, the eastern flanks of the Andes descend from alpine conditions through cloud forest into the lowland rainforest of the Amazon basin, while the western flanks descend to a coastal desert. High altitude, steep mountainous slopes, and extremes of temperature and rainfall would seem to provide an unpropitious environment for human habitation or crop domestication, but artificial selection applied to natural plant diversity beat the odds. When the Spaniards conquered Peru in 1535, there were at least 70 crops in cultivation, significantly more than in the Fertile Crescent or in any of the centers of domestication in Asia.

In the great worldwide dispersal of our species from Africa (chapter 3), the first people entered North America, probably along the coastal edge of the Bering Land Bridge from Asia, soon after the shoreline became free of ice around 16,000–17,000 years ago. Within 2,000 years of the opening of the land bridge, travelers along the Pacific coastal route had made it all the way to South America. Most of the coastline that they followed was later submerged by the rise in sea level that took place as the last ice age ended, but the remains of many coastal settlements farther inland have been found.

The earliest such settlement was at Monte Verde in south-central Chile. It was inhabited 14,600 years ago—a date that created disbelief among archaeologists when it was first discovered because it challenged the then-established view that North America was not settled until 11,000 years ago. The remains at Monte Verde tell us that the people who lived there ranged widely from seashore to mountainside for their food, hunting gomphotheres—a now-extinct elephant-like mammal—and palaeolama, also now extinct. The inhabitants collected seaweeds for food and medicinal use, just as people in the area do today, and they gathered about 50 species of plants, including wild potatoes (*Solanum maglia*) that were discovered stored in a pit. Wild potatoes grow at much higher elevations than Monte Verde, so they must have been fetched or traded from far away.

In the 2,000 years following their arrival, the lifestyle of the people living along the Pacific coast of South America gradually changed from the hunter-gatherer existence found in evidence at Monte Verde to a more settled pattern that increasingly depended upon horticulture. This change has been meticulously tracked through the archaeological record present at nearly 600 sites that have been excavated on the coastal plain and the western foothills of the Andes in northern Peru.

Seeds of a squash (*Cucurbita moschata*) recovered from an archaeological site indicate that this was the first vegetable to be cultivated in the area, about 10,500 BP. The seeds might have belonged to a wild squash, but those have a very bitter flesh that cannot be eaten, so the seeds more likely came from a domesticated plant. The earliest direct evidence of the Andean diet comes from 2,500 years later in starch grains found in dental calculus. Eight thousand years ago, the people inhabiting the Ñanchoc valley on the lower western slopes of the Peruvian Andes lived in many small settlements, eating peanuts, squash, beans, and manioc root (cassava). With the exception of squash, which we can assume was domesticated by then, none of these plants grew wild in this region, and so they must also have been cultivated. The peanuts, however, were small, resembling those of a wild species, which suggests that they were only recently domesticated. As a general rule, the seeds of domesticated plants like peanut, corn, sunflower, peas, and beans get bigger and bigger with time under artificial selection.

To the plant remains found in valley settlements add quinoa, an important grain crop that was domesticated in the Andean highlands, and cotton, a native of the coastal plains of Ecuador and northwestern Peru, to the list of plants cultivated there. The dental record shows that cultivated plants on the Ñanchoc menu were supplemented with wild gathered ones, particularly pacay (*Inga feuillei*), which is a tree that produces large edible pods containing a sweet white pulp. Tree fruit do not all contain starch, and of those that do, some contain starch grains that are not distinctive enough to be diagnostic of the species, so calculus does not provide a comprehensive list of all plants in the diet. To the inventory of plants that the calculus remains tell us were certainly eaten can be added others that we know were cultivated at a later date and would probably have been grown and eaten in the Ñanchoc valley. These include domesticated tree fruits such as guanabana, custard apple, guava, naranjilla, and lupini bean.

How archaeologists must delight in the poor oral hygiene and slovenly kitchen habits of our ancestors! Without dental calculus and the scraps of plant material trodden into the floors of ancient dwellings, we would not know that 8,000 years ago, farmers in the Ñanchoc valley enjoyed a varied and well-balanced diet and grew plants that had come from the four corners of South America: peanuts from the tropical south, cotton from the arid northwest, manioc from Amazonia, and quinoa from the high Andes. Neither would the extraordinary implications of this Pan-American cuisine be apparent to us. For all these crops to have converged in this valley means that they must have been domesticated in their disparate places of origin some time before. This in turn tells us that horticulture was widespread in South America at around the same time that farming was established in the Fertile Crescent.

The plant that is conspicuous by its absence from the menu in the Ñanchoc valley is the one Peruvian vegetable that is familiar to everyone outside Peru—the potato. The reason it was not grown in the Ñanchoc valley must surely be that this plant grows best in the cooler climate of higher altitudes in the Andes. Indeed, the potato's suitability to cool, wet climates was the key to its spectacular success in northern Europe after it was brought there. It is now the fourth most important staple in the world, after the cereals corn, wheat, and rice.

All the potatoes grown around the world belong to the domesticated species known as *Solanum tuberosum*, which descends from a single wild Andean species called *Solanum candolleanum* that was domesticated in the Andean uplands around Lake Titicaca at the boundary of Peru and Bolivia. However, in the Andes themselves, more than a hundred different wild potato species inhabit its valleys and mountains. This indigenous diversity is typical of mountainous regions because they present evolution with a complex range of different environments in which local adaptation is forged. Each valley contains a multitude of microclimates made by differences in altitude and aspect. Soils vary in moisture content from arid to soggy, and all these differences combine to create a plethora of unique places where natural selection can hone species to a local adaptedness that distinguishes the population in one place from that in another. Isolated plant populations, separated from each other in deep valleys between lofty ridges that pollinating insects cannot easily cross, diverge into separate species when left undisturbed for millions of years.

The indigenous people of the Andes domesticated not just one of the 107 wild potato species, but at least four of them, and it is estimated that there are 3,000 potato landraces still grown by indigenous farmers in South America today. One of the four domesticated species, called *Solanum hydrothermicum*, comes from an arid climate where other potatoes will scarcely grow. It could profitably be grown in other dry places around the world. By contrast, *S. ajanhuiri* is grown in a bitter cold and windy environment at an altitude of 3,800–4,100 meters around Lake Titicaca and will reliably produce a crop there in years when *S. tuberosum* fails.

Wild potatoes are useful as a source of genes that confer resistance to domesticated potato's many natural enemies. Species from hotter drier places in the Andes tend to be more resistant to leaf beetles, while those from cool or moist areas resist aphids better. A wild species called *Solanum berthaultii* has leaves that become as sticky as flypaper when an insect walks across them, breaking leaf hairs that release a resinous glue. Other wild potatoes have genes for resistance to late blight, a fungus-like disease caused by *Phytophthora infestans*. This is the pathogen that, compounded by poverty and overpopulation, caused the Irish potato famine in the 1840s in which a million people died and a million more fled Ireland. *Phytophthora infestans* has evolved resistance to the fungicides that have been used against it in modern times, and late blight continues to threaten potato production worldwide, as well as production of tomatoes and other plants in the nightshade family.

Like the progenitors of most vegetables, wild potatoes are poisonous and their domestication involved both selecting for reduced toxicity and devising processing methods to make them more edible. Regular non-toxic potatoes become toxic if exposed to light, which triggers the production of bitter-tasting poisons called glycoalkaloids. Fortunately, potato skin that has been exposed to light turns a warning green with chlorophyll, so such potatoes can be easily avoided or peeled to remove the toxic outer layer.

Only bitter potatoes will grow at altitudes of 4,000 meters and above in Peru, and these are traditionally processed into a freeze-dried, non-bitter product called *chuño*. To detoxify them, bitter potatoes are first exposed to freezing nighttime temperatures for several days, and then they are soaked for a month in a pit of water or in a streambed to leach out the glycoalkaloids. After that, they are freeze-dried for a night, then trampled underfoot to squeeze out water, and finally spread out to dry in the sun for 10–15 days. At the end of this process, the dried *chuño* can be stored indefinitely until needed.

The Incas stored enough *chuño* and dried, salted meat in their warehouses to supply the needs of the population for 3–7 years at a time, which provided their empire and their army with good food security despite a very variable climate and the natural disasters that this inevitably brings. When crops fail, the highlanders of Peru still fall back upon *chuño*. The Inca Empire stretched 4,000 kilometers along the Andes from southern Colombia to Santiago in Chile. When they came to power in about 1400 AD, the Incas built upon millennia of horticultural achievement by the various Andean peoples whom they conquered.

The Inca clearly understood that food is power and that the sun is its ultimate source. The founder of the dynasty Manco Cápac had declared that his father was the sun and his mother the moon. Using the agricultural surplus supplied by Andean agriculture to feed an army of stonemasons, Manco Cápac commanded that a Temple to the Sun be built in his capital, Cusco. When the Spaniards arrived, they found a huge wall made of massive interlocking stone blocks, decorated on the outside with friezes of solid gold and pierced by an entrance also sheathed in gold. Among the shrines found inside was a garden dedicated to the sun, furnished with life-size cornstalks of silver bearing corncobs of gold. Over the ground were scattered lumps of gold the size and shape of potatoes.

The Inca used their administrative talent and imperial power to purposefully spread agricultural technology and domesticated plants throughout their empire. Where there was a local rebellion against their rule, the Incas would forcibly move thousands of people, along with supplies of their local crops, to new locations where the inhabitants were loyal. Understanding that crops have specific environmental niches, the new locations were chosen by the Inca to enable the displaced people to farm their familiar crops in an environment similar to the one that they came from.

The effect of the Inca's imperial policy of pacification was to spread crops up and down the empire. The variety of vegetables grown gave both the food supply and the Inca Empire itself a resilience that nineteenth-century Ireland, dependent on just the potato, so woefully lacked. In addition to the four cultivated potato species grown in the Andes, almost 20 other root crops were domesticated and, though unknown outside Peru, many are still grown by indigenous farmers today. *Oca* (*Oxalis tuberosa*) is an exceptionally hardy plant whose wrinkled, stubby tubers are a staple for farmers living at altitudes above 3,000 meters in Peru and Bolivia. The tubers are brilliantly colored red, pink, yellow, and purple. Like several other Peruvian vegetables, there are both bitter and sweet varieties. The sweet roots can be eaten raw or cooked and when dried taste like figs. They were used as a sweetener before cane sugar, a domesticate originally from New Guinea, arrived shortly after the Spanish conquest. Bitter *oca* is freeze-dried and can be stored and eaten like *chuño*.

Yet another cold-hardy root crop that can be found in Andean markets is *ulluco* (*Ullucus tuberosus*), a tuber that comes in a waxy skin of many colors, including a candy-striped variety. Two flowers that are familiar to gardeners in temperate climates are grown as root vegetables in their native Peru: canna, or *achira* (*Canna edulis*), and nasturtium, or *mashua* (*Tropaeolum tuberosum*). To control pests, farmers grow *mashua* in mixtures with *ulluco*, *oca*, and bitter potato.

The evolution of another South American root vegetable illustrates why some varieties preserve their toxicity when they are domesticated. Manioc (*Manihot esculenta*) is a drought-tolerant crop that was domesticated on the southern edge of the Amazon basin where the climate is seasonally dry and lowland tropical rainforest gives way to savanna. The plant is a woody shrub in the spurge family that produces a large, starchy, tuberous root. Manioc is easy to grow in a tropical climate, thriving on acid soils that are poor in nutrients where other crops struggle. Though originating on the edge of the Amazon rainforest, it was widely grown in gardens by forest dwellers in the Amazon basin in pre-Columbian times.

The raw tuber deteriorates within days of being dug up, but if left in the ground, it is a dependable food source that can remain usable for up to two years. If you buy a manioc root in a store, it will come covered in wax that has been applied to preserve it. A major reason that manioc lasts so well in the ground is that the tuber consists of starch mixed with cyanogenic glycosides. These are molecules that liberate highly toxic cyanide when broken down by enzymes called glycosidases. These enzymes are released when plant cells are damaged by being chewed or crushed. So, as typically happens with plant poisons of all kinds, the chemical weapon is not released until it is needed. Cyanogenic glycosides are far from unique to manioc and occur in more than 2,500 other plants including common ones like bracken fern and white clover. The smell of bitter almonds is in fact cyanide, which is tolerable and even tasty when the dose is very low. However, manioc is the only staple food plant that can administer a fatal dose of cyanide.

Despite its toxicity, manioc is the staple food of more than 800 million people. The plant is known as cassava in Africa, where it was introduced 400 years ago and nearly half the population of sub-Saharan Africa depend upon it. To render manioc roots edible, they must be processed to remove the cyanide. Roasting or boiling does not detoxify this particular vegetable, but in fact makes it more dangerous because heat destroys the plant's own glycosidases, leaving the cyanogenic glycosides intact. If the manioc is then eaten, the cyanogenic glycosides release cyanide when they reach the intestines, where they react with glycosidases produced by bacteria there. Amazonian Indians detoxify manioc by grating the peeled tubers, releasing the cyanide into solution in the plants' juices that are then extracted by squeezing the flour in a woven sleeve press called a *tipiti*. Any remaining cyanide is then driven off by toasting the flour on a griddle.

The curious thing about manioc is that there are "sweet" nontoxic varieties as well as "bitter" toxic ones and that both types were derived from the same wild plants when they were domesticated more than 8,000 years ago. Why, when non-toxic varieties of manioc are available and bitter ones involve more work because of the processing required, do farmers grow the toxic ones at all? When farmers are asked this question, they give reasons that have to do with various aspects of food security. Bitter varieties are more productive, the tubers are less affected by pests, and they are less prone to theft by animals or people. Sweet as well as bitter varieties may both be grown, but the sweet ones are placed in gardens around the house where thieves can be warned away, while the bitter varieties can be safely left to fend for themselves in more distant plantings. Sweet varieties are also grown by communities who do not depend on manioc as a staple, but only as a secondary vegetable that can be substituted by something else if the crop is poor or stolen.

In the wild, the evolutionary relationship between plants and their natural enemies is like an arms race. On the one side, plants are under constant selection to improve their defenses, while in the enemy camp, on the other side, insects, fungi, and other plant consumers are under natural selection to overcome plant defenses in order to eat. This constant struggle has ancient origins. Fossils found in coal measures in Illinois show that 300 million years ago the tree ferns that dominated the swamp forests were under attack. Insects chewed leaves, pierced them to raid their sap, and bored through living stems and roots, just as modern insects do. There were even gall-making insects back then. These insects use a hypodermic-like ovipositor to inject their eggs into plant tissue. This act, or the presence of the egg, chemically stimulates the surrounding plant cells to proliferate into a mass, the gall, which provides food for the insect larvae inside and protects them from external attack.

Occasionally in evolutionary history, natural selection stumbles upon a key innovation that confers an advantage that has a dramatic effect upon fitness (the number of offspring contributed to future generations). These events are rare but epoch-making in their consequences because they unleash a multiplication of new species, all sharing the same advantageous innovation. If you have capers, radish, broccoli, cabbage, watercress, or arugula on the menu or any of the condiments mustard, wasabi, or horseradish on the table, then your meal benefits from a key innovation in the chemical warfare between plants and their natural enemies. This was the evolution of glucosinolates, which are produced almost exclusively by an order of plants called the Brassicales to which all of these food plants belong.

Glucosinolates, like cyanogenic glycosides, are another example of a two-component chemical defense. In fact, the biochemical pathway by which plants produce glucosinolates is similar to the one that produces cyanogenic glycosides and probably evolved from it. In the plant, the glucosinolate molecule and an enzyme called myrosinase are stored in separate compartments. When cells are damaged, the two compounds mix and the enzyme reacts with the glucosinolate molecule to release isothiocyanates, or "mustard oils." These compounds are toxic to many insects, nematodes, fungi, and bacteria, but have tumor-suppressing effects in mammals and are beneficial to human health.

The Brassicales evolved between 85 and 90 million years ago, and for a period of time these plants must have enjoyed a break from the attention of some of their enemies. But within 10 million years of the appearance of glucosinolates, a biochemical detoxification mechanism evolved among the pierid butterflies that enabled its larvae to feed on Brassicales without harm. This key innovation in the herbivore camp unleashed the exuberant evolution of more than a thousand new butterfly species, as insects carrying the genes that equipped them to feed on plants that were hitherto immune to attack spread and established themselves on any available member of the Brassicales.

This new group of butterflies became the subfamily Pierinae, better known as the whites, whose most notorious member, the cabbage white butterfly (*Pieris rapae*), is public enemy number one to any vegetable gardener. Caterpillars of this species can also tolerate cyanide. This is probably an evolutionary legacy from the ancestors of the Pierinae that fed on cyanide-producing plants before the transfer to Brassicales. Thus, the advantage in the chemical war between plants and their enemies switched to and fro as plants evolved new defenses from old ones and butterflies followed suit, evolving new detoxification mechanisms from old.

The glucosinolates are a chemically versatile class of defense compounds that have continued to evolve, particularly in the Brassicaceae, which is by far the largest family in the order Brassicales, comprising 3,700 species. Thale cress is a short-lived wild plant in this family whose genetics has been very intensively studied. A geographic survey of thale cress found that a gene that alters the chemical structure of its glucosinolates has two alleles (variants), which change in their relative frequencies from the south of Europe to the north. There is parallel geographic variation in the frequency of two specialist aphids that attack Brassicaceae, so investigators set out to test whether the change in glucosinolate type was caused by natural selection adapting the chemical defenses of thale cress for resistance to the most abundant local aphid species.

In order to do this, they set up experimental populations of thale cress containing 50:50 mixtures of the two glucosinolate variants and exposed the populations to either one aphid or the other for five generations. At the end of the experiment, the frequency of glucosinolate types had diverged between the two aphid treatments. The populations exposed to five generations of selection by the aphid that is common in northern Europe ended up with a high frequency of the glucosinolate that is common in the north, while the populations exposed to the aphid more common in the south had a high frequency of the glucosinolate that is common there. These experimental results strongly support the hypothesis that the geographic variation in glucosinolates reflects local adaptation to prevalent natural enemies.

The ever-running evolutionary battle between organisms and their natural enemies has been compared to the situation of the Red Queen in Lewis Carroll's *Through the Looking-Glass and What Alice Found There*. In the story, Alice realized that even though she ran as fast as she could, in the looking-glass world, this did not get her anywhere. The Red Queen then explains to Alice: "Now, *here*, you see, it takes all the running you can do, to keep in the same place." In evolutionary biology, the Red Queen's hypothesis is the idea that the evolutionary arms race between organisms and their natural enemies means that both parties must continually evolve just to avoid extinction.

Continual evolution can only happen if there is a ready supply of genetic variation from which natural selection is able to fashion new weapons and new defenses. Plants that reproduce entirely vegetatively, as the potato does when it is grown generation after generation by replanting tubers, become genetically uniform, and it is then only a matter of time before they are wiped out by their natural enemies. Manioc is also grown and replanted vegetatively by rooting pieces of stem. The way out of this evolutionary dead end is through sexual reproduction. Sex creates new gene combinations among offspring, so that they are different from each other and from their parents.

Although the vegetable is propagated vegetatively by growers, potatoes do reproduce sexually, and the stray seedlings from these uncontrolled unions were the source of new varieties before deliberate potato breeding was begun. The same is true of manioc, and in this crop it has been discovered that farmers, in preferring the biggest seedlings that appear, unwittingly select for the new plants that contain the greatest genetic variation because these are also the ones that grow better.

Sexual reproduction not only maintains genetic variation and thereby reduces the risk to crops from disease epidemics, but it also makes hybridization between different species possible. Many crops are of hybrid origin, including bread wheat (chapter 4) and a lot of vegetables. There are six vegetable species in the genus *Brassica*, but rather strangely they all have a different number of chromosomes, ranging from just 16 in black mustard (*Brassica nigra*) to 38 in canola (*B. napus*). Such variation is usually the result of hybridization between plants that themselves have different chromosome numbers. Finding what combinations of parents created the hybrids is like trying to solve a Sudoku puzzle. Perhaps, then, it is no accident that in 1935 the *Brassica* Sudoku was solved by a Japanese botanist.

The botanist, whose name in English is simply U, found that when he drew a diagram that arranged the three species with the lowest chromosome numbers at the corners of a triangle, the three remaining species all fell neatly into place as pairwise combinations between the corners. So, hybridization between wild cabbage (18 chromosomes) and wild turnip (20 chromosomes) produced canola (38 chromosomes). Black mustard (16 chromosomes) and wild cabbage hybridized to produce Ethiopian mustard (16 + 18 = 34 chromosomes), and black mustard and wild turnip produced brown, or Indian, mustard (16 + 20 = 36 chromosomes).

Modern genomic analysis has dated the origin of the species in U's triangle and pinned some of the events to locations on the map. The common ancestor of all the brassicas evolved some 24 Ma in North Africa. The three species at the corners of the triangle then evolved in different places. Black mustard evolved 18 Ma in the western region of North Africa and from there spread into southwest Asia, where it gave rise to the common ancestor of wild cabbage and turnip 7.9 Ma. This common ancestor split 2.54 Ma, evolving into wild cabbage in the western part of its range in the Mediterranean, while farther east it evolved into wild turnip, reaching central Asia around 2 Ma.

The three hybrid brassicas in U's triangle each formed when their respective parents were brought into contact, as a direct or indirect consequence of agriculture. For example, Ethiopian mustard is thought to be the result of a cross between wild black mustard weeds and cultivated cabbages with which they were growing. In addition to the many vegetables already mentioned as originating from the artificial selection and domestication of wild cabbage (*Brassica oleracea*), *B. rapa* produced turnips and Chinese cabbage, and *B. napus* produced the oilseed crop canola (oilseed rape) and rutabaga (Swede).

Despite all their wonderful variety, we eat vegetables for a single overriding reason: their nutritional properties—especially the carbohydrates they provide. To make vegetables edible, we have lowered the natural defenses of these plants by artificial selection and by cooking and processing them. Perhaps it is ironic, then, that when cooking we like to add other plants to the pot for the sake of their defensive compounds. Potato salad tastes so much better with chives, tomatoes benefit from basil, the flavor of peas is given a lift by mint, and the culinary uses of garlic are too numerous to mention. And, like the pierid butterflies to whom the whiff of a brassica's poisonous glucosinolates is the smell of dinner, our own species has crossed the planet in search of spice plants whose chemical weapons lure us to the table.

# 9

# *Herbs and Spices— Piquancy*

Corn, cabbages, cows, and cauliflowers are all witness to how we as a species have shaped nature by applying the evolutionary force of domestication. In the last 10,000 years, we have remixed, reorganized, and multiplied genomes, rearranged genes, fattened animals, and made everything in the produce market bigger and tastier. These are as much the achievements of art as of science, since only in the last 100 years have we begun to understand the genetics of selection sufficiently to use this knowledge. Whether through art or science, there is no doubt that we have redrawn and remade the view from the kitchen window. Even in the depths of the Amazon rainforest, an Amerindian kitchen garden is arrayed with manioc, corn, beans, sweet potato, and fruit that are the products of indigenous domestication. Humans, then, are in charge of edible nature. Or are we? If there is a case to be made that we are not, and that the tables are turned upon us by our appetites, then a good foundation for the argument could be built upon the seductiveness of spices.

Herbs are the near-at-hand plants with aromatic leaves that we can grow ourselves and use fresh by the handful. Spices are the pungent seeds, resins, bark, and other plant parts that until modern times were rare and exotic. Passing through many hands on their journey from east to west across the globe, their sources were unknown lands sketched by imagination on maps drawn in ignorance. Mystery was an adjuvant to odor in the allure of cloves, ginger, pepper, cinnamon, mace, and nutmeg. The Greek historian Herodotus wrote:

> The Arabians say that large birds bring those dry sticks called cinnamon for their nests, which are built with clay on precipitous mountains that no man can scale. To surmount this difficulty, they have invented the following artifice: having cut up into large pieces the limbs of dead oxen and other beasts of burden, they lay them near the nests and retire to a distance. The birds fly down and carry off the joints to their nests, which are not strong enough to support the weight of the meat and fall to the ground. Then the men come up and gather the cinnamon, and in this manner it reaches other countries.

Perhaps this story, as good as any invention in *The Arabian Nights*, was the result of Chinese whispers transmitted from ear to ear as cinnamon was passed from hand to hand along the spice route from Asia, confusing and embellishing a tale that started with a kernel of truth. Edible birds' nests harvested from cave walls in Borneo have been used in eastern cuisine for centuries, but these are made of the dried saliva of two species of swiftlet and not from cinnamon sticks, which come from the bark of a tree that is native to Sri Lanka.

Spices were used in medicine as well as in the kitchen. They were so rare and their value so high that the search for the sources of these commodities, as well as the desire for gold, provided the motive for voyages into the unknown by Christopher Columbus and Ferdinand Magellan. Both the "discovery" of America by Columbus and the first circumnavigation of the globe by Magellan were incidental to their quest for spices. Hernán Cortés, conqueror of the Aztecs, promised the Spanish king who sponsored his voyage that he would discover the spice islands of the East, or "Your Majesty may punish me as one who does not tell the truth." Taking a westerly route to reach the Indies was a gamble that never paid off in spices. Chili, the quintessential spice of Mexico and until then unknown in Europe, never traded for the prices fetched by the spices of the East or matched the value of American gold and silver.

The trade in spices from east to west began more than 3,000 years before the merchants of Europe decided to seek the source and control the market for themselves. The mummy of Pharaoh Ramses II, interred circa 1213 BCE, was embalmed with black peppercorns in its abdomen and nasal cavities. The black pepper vine is endemic to the wet forests of South India, whence it was probably collected by hunter-gatherers and traded to the west coast, where eager buyers, arriving by boat, would have been ready to carry it across the Indian Ocean. The overland pepper route from the forests on the east coast of India and across the continent to the west coast seaports was certainly well-established by Roman times, as attested by the trail of lost Roman coins it left behind. The sea crossing from India was also one of the routes by which chickens arrived in Africa (chapter 7). Cinnamon is a spice mentioned in the Old Testament and must have been regularly carried to the Levant in a similar way. By 1100 BCE the cinnamon supply was sufficient that the Phoenicians traded small sealed flasks of extract around the Mediterranean.

Of the other classic eastern spices, ginger probably came from northeastern India or southern China, though its exact origin is unknown because no wild relative has yet been identified. Cloves, nutmeg, and mace were the rarest and most sought-after spices and came from the remotest source. Cloves are the dried flower buds of a small tree that grew only on a few islands in the North Moluccas of Indonesia. Nutmeg and mace also came from Indonesia, originally being found on only a handful of islands in the Banda group. Nutmeg trees bear peach-like fruit that split open when ripe to reveal a seed (the nutmeg) that is wrapped in a bright red aril. When dried and separated, the aril turns orange-brown and becomes mace.

Spices and herbs all have anti-microbial properties, which, some have argued, might explain their heavy use in hot countries where meat is prone to rot very quickly. The spiciest cuisine is to be found in tropical and subtropical regions, perhaps, the argument goes, because spices are essential there to making meat safe and palatable. Compare the hot cuisines of Louisiana or New Mexico with the traditionally milder tastes of Seattle or Boston, for example. Hotter recipes are used in the south of India than in the cuisine of the north, and the same pattern is seen in China. In a Chinese restaurant, the hottest dishes on the menu, like kung pao chicken, are from Sichuan in southwest China. Unfortunately for the theory that spices are used to make bad meat palatable, they are not actually very good for this purpose and with their strong tastes can actually make matters worse. Furthermore, salting, drying, smoking, and fermentation are all much better methods of food preservation, and all are very widely used. As for the correlation between climate and spice use, as Mark Twain once famously said, "There is something fascinating about science. One gets such wholesale returns of conjecture out of such a trifling investment of fact." The correlation might simply be explained by the geographical availability of spices, many of which tend to be tropical in origin.

Garlic and onion are two ingredients that have very strong antimicrobial properties, but they are neither used for food preservation nor are they tropical in origin. These two wonderful plants and more than a dozen kitchen relatives including leeks and chives all belong to the genus *Allium* that contains about 500 species. All of them are chemically defended by sulfur-containing compounds that are the basis of both the pleasure and the pain of the onion. An intact onion bulb or garlic clove has no smell because, as with glucosinolates in brassicas and cyanide in manioc, the chemical arsenal consists of two components that do not become noxious until they are mixed and can react with each other. Cutting or crushing an *Allium* releases the two components, a precursor and an enzyme, from their separate compartments in the cell. Crushing garlic starts the transformation of the garlic precursor called alliin into the molecule called allicin, which is the active principle in garlic. A similar precursor in onion reacts first with the same enzyme found in garlic, but then there is a second reaction caused by another enzyme that produces the molecule that makes grown men cry.

Plants produce tens of thousands of different chemical compounds that seem to serve only, or mainly, as defenses against natural enemies. These compounds are the active ingredients of herbs and spices as well as medicines like quinine and aspirin, narcotics such as opium and cannabis, and the daily fixes of coffee and tea. Ever bountiful with its gifts but always parsimonious in its means, evolution has generated this chemical diversity in plants by ringing the changes on a limited number of components. The diversity of molecules is made inside a plant cell by a handful of basic biochemical pathways that branch to many chemical destinations. Each pathway begins by making building blocks with fixed numbers of carbon atoms. For example, the terpenoid pathway that makes the aromatic compounds in many spices and herbs begins with a basic building block of five carbon atoms. Like building with LEGO, the basic five-carbon units are then joined together to form bigger chains, or skeletons, of various sizes and configurations. A class of terpenoids with a ten-carbon skeleton called the monoterpenes provides plants in the mint family—such as basil, thyme, oregano, and rosemary—with their characteristic aromas. At the other end of the scale, natural rubber is a terpenoid with a giant carbon skeleton of up to 100,000 five-carbon units, or half a million carbon atoms altogether.

In the second phase of construction, the multi-unit carbon skeleton is tailored with additions and rearrangements. The two phases of construction—the first making a range of carbon skeletons and the second embellishing these in a variety of ways—are capable of making a huge number of different molecules. More than 40,000 chemical products of the terpenoid pathway alone are known. Plants invariably make not just one, but many aromatic molecules, producing chemical diversity both within individuals and from plant to plant. Thus, any reasonably well-stocked garden center will offer mint varieties that smell of lemon, apple, geranium, ginger, peppermint, spearmint, and so on. Each of these scents is the result of a different mixture of monoterpenes, but because of the way biochemical pathways branch, small genetic changes can produce quite different mixtures and aromas in different plants. Only a single gene affecting one enzyme is responsible for the difference between peppermint and spearmint aromas, but its effect is like pulling a lever that switches points on a railroad track. One allele leads in the direction of the peppermint mixture of monoterpenes, the other to the spearmint mixture.

Why do herbs like mint produce such a variety of defensive compounds, when one might think that natural selection ought to favor the production of a single, superlatively deadly monoterpene? A fundamental reason is that natural selection makes incremental improvements by tinkering with existing mechanisms, so natural enemies only ever have to overcome small changes in plant chemistry, and they experience strong natural selection to do so. Hence, the gradual nature of evolution does not allow a plant ever to evolve a final killer blow against its enemies. Even the evolution of a new class of poisons such as the glucosinolates provided plants in the Brassicales with only a temporary respite from their natural enemies (chapter 8).

A second reason for chemical diversity is that when facing an array of natural enemies that are all evolving, a flexible strategy such as that provided by a battery of defenses is of great advantage. This point is illustrated by the way in which the genetic difference between spearmint and peppermint was discovered by scientists who were screening varieties of spearmint to find one that was resistant to a fungal disease that was affecting commercial production of the plant in the United States. A plant that they found to have higher resistance to the disease turned out to smell like peppermint. The difference in monoterpenes between spearmint- and peppermint-smelling plants was the source of the disease resistance. A diversity of chemical defenses is an advantage against an evolving enemy or a field of many foes.

Chemical defenses also vary because environments differ, requiring local adaptation. In the Mediterranean climate of the South of France, wild thyme occurs in six different forms, each one characterized by a different dominant monoterpene molecule. Genetic analysis of the six different forms, which are called chemotypes, found that the chemical differences between them were due to five genes. Each of the five genes controls one step in the biochemical pathway leading ultimately to the production of the monoterpene thymol, which has the characteristic aroma of thyme. A dominant allele at the genetic locus controlling the first step in the pathway curtails it at that point, producing the monoterpene geraniol that has a lemon fragrance. Further along the pathway, a gene controlling the third step determines whether the monoterpenes produced will have a phenolic or a non-phenolic structure. Only chemotypes with pathways that pass this point produce phenolic monoterpenes and smell of thyme.

Scientists studying populations of wild thyme in the South of France found that near the city of Montpellier, around a village called St. Martin-de-Londres, chemotypes were distributed in a very distinctive pattern. St. Martin-de-Londres lies in a basin surrounded by mountains, and none of the thyme growing near the village actually had the characteristic smell of thyme. In fact, all the chemotypes growing below an elevation of 250 meters turned out to be non-phenolic. By contrast, all the plants growing above the 250-meter contour were of phenolic chemotypes and smelled of thyme.

The explanation for this strange distribution of chemotypes turned out to be the difference in winter temperatures between the floor of the basin near the village and the mountainsides around it. In cold winters, a thermal inversion traps cold air, which is denser than warm air and therefore sinks beneath it, in the basin around St. Martin-de-Londres. The mountainsides above 250 meters where phenolic chemotypes grow are in a zone of warmer air and escape the coldest weather. Experimental transplants of chemotypes between the two areas, above and below 250 meters, found that phenolic chemotypes are killed by early winter cold, which in some years could reach well below −15°C. By contrast, in places with warmer winters, phenolic chemotypes survive drought, resist insect enemies, and grow better than non-phenolics.

A twist to the tale confirms the importance of local adaptation to cold winters. Chemotypes around St. Martin-de-Londres were first recorded in the 1970s when very cold winters were common, but climate warming since that time has meant that no winter since 1988 has been as cold as previously. A resurvey of the distribution of chemotypes in 2010 found that phenolic plants have begun to colonize the basin where, back in the 1970s, there were none.

The Mediterranean region is rich in plants of the mint family, among which it is a general rule that the highest yields of essential oils containing phenolic monoterpenes are produced in the hottest places. The mixture of monoterpenes also varies geographically. Rosemary contains four or five major monoterpenes: in France and Spain, rosemary essential oil is dominated by camphor; in Greece, eucalyptol is dominant; and in Corsica, the oil is almost entirely verbenone. Why there are these regional differences is not known.

So far we have only considered half the evolutionary story of herbs and spices—the botanical half, but of course the reason we are interested in these plants in the first place is the effect that they have upon our senses. From an evolutionary point of view, it is surely puzzling that the plant chemicals that deter and poison most animals have exactly the opposite effect upon us. This paradox only deepens when we see how these alluring anti-feedants are sensed. The aromas of herbs and spices stimulate olfactory receptors that, operating in concert with one another, help the brain distinguish nice from nasty (chapter 6). In addition, several herbs and most spices also stimulate pain sensors on nerve cells called nociceptors. Nociceptors are found in all pain-sensitive parts of the body. The ones in the face, eyes, nose, and mouth transmit signals to the brain via branches of the trigeminal nerve. Nociceptors are equipped with a range of receptors called TRPs that react to external stimulation by generating a nerve impulse. Each TRP type is activated by a range of different stimuli, such as heat, cold, pressure, and certain chemicals.

It is because TRPs react to physical stimuli such as heat and cold, as well as to chemicals, that we experience some spices as "hot" and certain others as "cool." Chili peppers can feel like they have set your mouth on fire because the active ingredient of chili is a molecule called capsaicin that stimulates the receptor TRPV1, which also senses heat. Likewise, menthol, a monoterpene produced by mint, creates a cooling sensation because it triggers receptor TRPM8, which senses cold.

Other herbs and spices trigger various TRP receptors, which in concert with olfactory receptors lend each its characteristic flavor. Like chili, black pepper and Sichuan pepper both stimulate TRPV1, but Sichuan pepper also triggers two other receptors, TRPA1 and KCNK, which both produce a tingling sensation that is highly characteristic of this cuisine. When I first tried it in a restaurant in London's Chinatown, the cook used so much Sichuan pepper that my mouth went entirely numb. I should have taken this as the warning that nature intended it to be, because the restaurant then stung us by overcharging on the bill as well.

The pungent principles of mustard, wasabi, and horseradish, as well as the different ones of garlic and ginger, hit TRPA1 hard and soft pedal on TRPV1. Monoterpenes in thyme and oregano hit TRPA3 and soft-pedal on TRPA1. Cinnamon only stimulates TRPA1, but lemongrass attacks the four receptors TRPM8, TRPV1, TRPA1, and TRPV3 in that order of diminishing strength. Thus, the flavor sensations of herbs and spices are created in the brain by different combinations of signals sent by olfactory receptors in the nose and nociceptors in the tongue and mouth.

If you have ever touched a sensitive part of your anatomy after handling chili, you will know that it is not only the nociceptors in the mouth that are equipped with TRPV1 receptors. This is also why very chili-hot food burns on the way out as well as on the way in. Plants are not the only organisms to target TRP receptors in order to inflict pain on their attackers. Toxins found in tarantula venom target TRPV1, too.

The TRP receptors are an evolutionary ancient system that we share in its essentials, not only with other backboned animals, but also with insects, nematode worms, and even yeast. This explains why so many receptors that plants target to hack into the pain-sensing circuits of herbivores also work on our senses. But why do we respond positively to substances that activate pain receptors and that cause aversion in other species? The answer is that initially, on first encounter with pure chili and all the other TRP-triggering spices and herbs, the usual reaction is indeed aversive. We do react as you would expect to substances that trigger nociceptors when we have not tasted them before. A liking for these substances, which of course are not to everyone's taste anyway, is acquired. This is true of bitter-tasting foods too (chapter 5).

How is it that we can acquire a taste for aversive chemicals? The receptors that alert us to potential poisons are only the first line of defense against hurt. If the chemicals turn out not to be poisonous after all, then we can learn to enjoy the stimulation rather than shun it. This is advantageous and favored by natural selection because plants contain a lot of nutrition that we would needlessly forgo if we just took the plant's chemical word for it when it signals "I'm poisonous—don't eat me!" The fundamental explanation is in the dose. A small insect taking a large bite out of a poisonous plant is exposed to a much bigger dose per unit of body weight than a large animal like us taking a small bite of the same plant. Hence, what is poisonous to an insect eating a thyme leaf can become flavorsome to us when added in small amounts to our food. However, it is possible for us to overdose on some spices—nutmeg poisoning is not unknown.

Although TRP receptors have an ancient origin—like the taste receptors discussed in chapter 5—there has been plenty of evolutionary change resulting in differences in sensitivity between species. Certain TRP genes have been lost in some species, while others have changed function. For example, the cold-sensitive receptor TRPM8 has been lost in certain fish. The TRPV1 receptor that in mammals like us is so sensitive to capsaicin is insensitive to this chemical in birds, where it hardly raises a "cheep."

Chili plants use the different sensitivity to capsaicin of mammals and birds to their advantage. Experiments with wild chili plants in southern Arizona found that birds would take the ripe fruits and void the seeds in a germinable condition, but that rodents would not touch them. Rodents that had not previously encountered hot chilis would eat a variety that lacked the ability to make capsaicin, but the seeds in their feces were broken into fragments and could not germinate. Thus, capsaicin is a selective deterrent that prevents rodents from eating and destroying the seeds of the chili plant, while not scaring away birds that take the fruit and safely disperse its seeds.

Capsaicin is unique to the chili genus, *Capsicum*, but not all chili species are hot, and pungency varies a great deal, even within species. For example, domesticated varieties of *Capsicum annuum* range from the entirely innocuous bell pepper to fiery-hot chili peppers. The presence or absence of capsaicin depends upon a single gene called *Pun1*, but other genes plus growing conditions influence just how hot a chili plant that is capable of producing capsaicin actually is.

Just as wild populations of thyme vary in the frequency of plants that produce phenolic monoterpenes, so some wild chili populations also vary in the frequency of plants that produce capsaicin. And, as in thyme, this variation is due to adaptation to local conditions. A study of the wild chili species *Capsicum chacoense* in Bolivia, where chilis probably first evolved, found that populations were polymorphic—meaning that some plants were hot and some were not. Capsaicin in the seeds of hot plants protected them from a fungus called *Fusarium*. This fungus was most frequently encountered in wet environments where a bug lives that punctures chili fruits, providing an entry point for the fungus. In this kind of environment, capsaicin-producing plants were at an advantage and were in the majority. In drier places where the chili grew but the bug was absent and fungal infection was low, non-pungent plants were in the majority.

Since capsaicin protects seeds from rodent predators as well as from fungal infection, one would expect pungent plants to be at an advantage in any environment where rodents are present, not just ones where there is a risk of fungal infection. So, why were plants growing in drier places non-pungent? The answer turned out to be that pungent plants of this species did not grow as well as non-pungent ones in drought conditions. In fact, pungent plants produced half the number of seeds that non-pungent plants produced when water was scarce, though this difference was not seen when water was more available. This study shows that the chemicals that plants make to protect themselves do not come without a cost, which in this case was paid in the currency of seeds. Evolution balances costs and benefits as these are shaped by numerous ecological factors, such as here where attacks by bugs on fruit and on seeds by fungi and rodents, as well as the availability of soil moisture all come into play.

Herbs and spices illustrate the complexity, unpredictability, and even the irony of evolution. These plants have been armed by natural selection to deter the animals that would call them lunch, but we relish them for their very poisons and add them freely to our dinner. If spices suggest that sometimes we can be slaves to our senses, then dessert is the commonest weakness and the cheapest luxury.

# 10

## Desserts—Indulgence

Mozart, the genius from whom music poured forth like sweet wine, was—according to modern Renaissance man, opera director, and gourmet Fred Plotkin—fueled by Viennese cakes and pastries. Vienna has to be the capital of patisserie and the ultimate destination for all who indulge in desserts. This is the home of *Apfelstrudel*—a delicious pastry made with a sweetened apple filling, flavored with cinnamon, and baked inside a parcel made of sheets of the most membranous phyllo, then brushed with butter and dusted with confectioner's sugar. It is the city where two establishments fought a seven-year legal battle over which had the right to claim to be the maker of the original *Sachertorte*, a classic Viennese chocolate cake to die for.

A great deal of culinary art and imagination goes into creating desserts, but despite the variety of flavorings used and the effort that goes into their preparation, desserts have only three basic ingredients: carbohydrates (sugar and starch), fat, and ingenuity. Take the Baked Alaska, for example. This is ice cream baked inside an insulating shell of meringue, producing the startling juxtaposition of oven hot and ice cold in a single dish. Baked Alaska has an even more ingenious inverse called Frozen Florida, invented by Nicholas Kurti (1908–1998), who was a low-temperature physicist and one of the founding fathers of molecular gastronomy. The recipe for Frozen Florida takes advantage of the fact that frozen water is permeable to microwaves, so by using a microwave oven it is possible to heat jelly inside a frozen casing of ice cream. Despite all the ingenuity of these dessert inventions, the former is basically fat encased in sugar and the latter is sugar encased in fat. This is, of course, a most inadequate and unhelpful way to describe a dessert in a recipe or cookbook, but it does get to the evolutionary essence of what desserts are all about—calories.

You do not have to delve too deeply into the evolution of human impulses to understand why we love carbs and fat so much: they are, after all, pure sources of energy for which we have specific taste receptors (chapter 5). The sweet-taste receptor in our taste buds senses sugar in sweet foods and also the glucose that is released from starchy foods by the enzyme α-amylase in saliva. Chemically, glucose, sucrose, and other sugars are described as simple carbohydrates, while starch is a complex carbohydrate that is a polymer of glucose. The distinction between simple and complex carbohydrates is nutritionally important, as we shall see. Saliva also contains lipase enzymes that break up fats, releasing fatty acids that stimulate their own sensors in our taste buds. Thus, we are well-equipped by evolution to detect both the kinds of high-energy food that we so relish.

The sugar glucose is a universal biological fuel that powers everything living. Plants, insects, yeast, and humans all trade or purloin this biological fuel. It is carried in solution through the blood vessels of animals and is transported around plants in the form of sucrose—a sugar molecule that is one part glucose and one part fructose. This is the sweet liquid that the Canadian farmer collects from her stand of sugar maples when the sap rises in spring. The sap contains only about 2 percent sugar and has to be boiled to concentrate the sugar and flavors in maple syrup. By contrast, sugarcane, which is a tropical grass, has sap that contains 20 percent sugar. This plant was domesticated in New Guinea, possibly 8,000 years ago, and is now grown all over the Tropics. The sap is so sweet that the traditional way to use the plant is simply to peel the stem and to chew the inner pith.

The sugars in nectar are the main reward that draws honeybees and other pollinating insects to flowers. Nectar-feeding insects ferry sugar from its solar-powered source in plants across a cable-less energy grid that can extend for thousands of meters, connecting all manner of animals, including ourselves, into this primary font of calories. Honeybees turn nectar into honey by reducing its water content and increasing the concentration of sugar to more than 80 percent, a point where yeast—a sugar-stealing fiend— cannot ferment it. High concentrations of sugar are preservative. This is why jellies, jams, and candied fruit, as well as honey, can be safely stored without refrigeration.

Besides being used as a fuel, glucose, especially in plants, is also used as a source of carbon atoms from which structural compounds like cellulose are built. Chemically, there is very little difference between cotton candy, which is spun sugar, and cotton wool, which is pure cellulose—a polymer built from glucose. For us, the first is a food and the second is indigestible, but there are organisms with exactly the reverse relationship to these two forms of sugar. You might imagine that cows and other animals that live exclusively on plants can digest cellulose, but in fact no animals are equipped with the enzymes to do this and all depend upon microorganisms in their guts to do the job for them. For these bacteria, cellulose is appetizer, main course, and dessert all rolled into one.

Honey is certainly the most ancient food on the dessert menu. Orangutans and chimpanzees probe bee nests with sticks to extract honey, and they also eat bee larvae, garnishing their haul of sweet stuff with protein. Since our great ape cousins eat honey, this may well have been a part of the hominin diet since before our ancestors and those of chimps parted company more than 5 million years ago. This is speculation, of course, and it is not until the Paleolithic that we have direct evidence of a honey habit. Twenty-five thousand years ago, the artists who decorated the famous Altamira cave in Spain with paintings of large beasts thundering across the mammoth steppe also depicted bees, honeycomb, and honey-collecting ladders in a smaller side cave, appended like a small helping of dessert to the main meal of wild aurochs.

Similar pictures of honey collecting appear in Paleolithic cave art in many parts of the world, but examples are most common in Africa, where the diets of present-day hunter-gatherers indicate just how important honey can be to this form of subsistence. For two months of the year in the wet season, the Efé, who inhabit the

Ituri Forest in the Democratic Republic of the Congo, live almost entirely upon honey, bee larvae, and pollen, each consuming the equivalent of about three average-size jars of honey a day. The less intense, more year-round consumption of honey by the Hadza of Tanzania is probably more typical of hunter-gatherers. They live in a grassland savanna landscape that is peppered with baobab trees, whose trunks and limbs afford cavities for nesting honeybees. The Hadza obtain 15 percent of the calories in their diet from honey. Hadza and other African hunter-gatherers are helped to find honeybee nests by a remarkable symbiotic relationship with a bird called the honeyguide, which has the transparently descriptive scientific name *Indicator indicator*.

Honeyguides eat insects including, when they can get them, bee larvae and beeswax. They do not consume honey, but they have been seen to search for bee nests and even, in the cool of the morning when honeybees are too torpid to sting, poke their heads inside a nest entrance, apparently to check for activity. Honeyguides are unable to access the inside of honeybee nests on their own, but instead they recruit human help by flying into Hadza camps and uttering a characteristic call that is recognized by people there as an invitation to follow. Hadza may also summon honeyguides with a call of their own that can bring the birds to them from as much as a kilometer away.

The honeyguide-human relationship was first recorded in the seventeenth century, and in more recent times the anecdotes of that period were regarded as a romantic myth. However, scientific studies have discovered, just as African hunter-gatherers themselves claimed, that honeyguides and people do communicate and work together in pursuit of honey. When a honeybee nest has been found in a baobab, a Hadza man will use an ax to sharpen wooden pegs and then drive them into the branchless lower trunk of the tree to form a ladder, by means of which he can reach the bees' nest. A burning brand is then used to pacify the nest with smoke, just as beekeepers do, and the nest is excavated from the trunk with the ax. The honeyguide-human relationship is mutually beneficial. By following honeyguides, Hadza honey hunters can find a honeybee nest in less than a fifth of the time that it takes without a bird's help. Moreover, the nests that honeyguides find are much bigger and contain a lot more honey than those found by hunters searching on their own. For their part, honeyguides gain access to food resources that would otherwise be unobtainable.

How did the honeyguide-human relationship evolve? One idea is that honeyguides evolved their guiding behavior in cooperation with another species, such as the honey badger, which is a generalist carnivore that does occasionally raid bee nests, and that this behavior then transferred to humans. Plausible though this hypothesis may sound, scientific observers of honeyguides have never actually seen the birds guide any species but humans. Thus it seems at least possible, perhaps even likely, that the symbiosis is an ancient one, maybe older than our species itself.

Control of fire is essential to the success of the bird-human relationship because smoke is used to pacify the honeybees, so the symbiosis could have evolved in the time of our ancestors *Homo erectus*, whom it is believed used fire in cooking. It has even been suggested that the honeyguide-human relationship could be older still if earlier hominins used the repellent and protective properties of herbs to pacify bees and to relieve the pain of bee stings, as some human societies around the world do today. However long it has been that a love of sweet things has tempted us to brave painful and life-threatening bee stings, there can be no doubt that honey theft by animals must have been what originally drove the evolution of the sting in honeybees. There are many species of stingless bees, but they have smaller nests stocked with less honey, or even none at all.

Just as honeybees protect their coveted calorific resources, so, too, do plants protect their nectar from robbers that would take the reward without delivering the service of pollination in exchange. Thus, in many flowers, natural selection has hidden nectaries at the bottom of long tubes that can only be reached by faithful pollinators equipped by evolution with a sufficiently long proboscis. In other cases, the nectar is laced with poison. It is not clear how, or even whether, the toxins in such nectar protect it from robbers, but the poison is selective in its effects in a way that suggests that it might serve that function. Honeybees are not deterred by toxic nectar, while humans who consume the honey made from it become seriously ill. Thus toxic nectar could potentially deter browsing mammals from eating the flowers without discouraging visits from pollinating insects. Toxic nectar is produced by a number of *Rhododendron* species including *R. ponticum*, by oleander *Nerium oleander*, and by mountain-laurel *Kalmia latifolia*.

The ancient Greek geographer Strabo was a native of the region near the Black Sea, now in modern Turkey, called Pontus, after which the poisonous *Rhododendron ponticum* is named. He tells the story of how the people of Pontus, being quite familiar with the poisonous potential of the local honey produced when the rhododendrons were in bloom, overcame an army led by the Roman general Pompey by strewing the route of his marauding troops with poisonous honeycomb. Three squadrons were incapacitated by eating the sweet bait and were then slaughtered to a man.

Such is the wholesome image and unalloyed pleasure of a spoonful of honey that the very idea that it might sometimes be toxic has in modern times attracted incredulity. The author of an entry in the 1929 edition of *Encyclopaedia Britannica* derided the account of the ancient Roman author Pliny the Elder, who wrote in his *Natural History* of "mad honey" from the region of the Black Sea. Pliny correctly attributed the neurotoxic effects of mad honey to the nectar of rhododendron, azalea, and oleander, all of which were known to have toxic leaves, but the *Encyclopaedia Britannica* preferred to imagine that "in all likelihood the symptoms described by these old writers were due to overeating." Cases of poisoning by mad honey still occasionally occur in Turkey, most often among middle-aged men who consume it quite deliberately in the vain hope that it will revive their flagging sexual prowess.

If, in the marketplace of nature, sugary sap is a liquid currency that may be transported, stolen, saved, or spent, then fat is money in the bank, stored up close and personal inside the body for when it is needed. Ounce for ounce, the fat in butter contains more than twice the calories of sugar. As an ingredient, fat appears in most dishes. A recipe for a delicious dessert that contains no fat whatsoever is a rare thing. This is not just because fat is tasty in its own right, but also because many flavor molecules are fat-soluble and so fat is needed to deliver them to our olfactory receptors.

Fats come in various guises, including those made by plants as an energy store to supply their seeds. The melt-in-the-mouth deliciousness of chocolate is due to the happy coincidence in the seeds of *Theobroma cacao* of a large helping of a fat that melts at body temperature and the alkaloid theobromine that acts as a stimulant. Add sugar to the mix and is it any wonder that the result can be almost addictive? Desserts are not by themselves responsible for excess consumption of calories, but a calorie-bomb cake does represent a good example of why being over-weight or obese is a major issue in public health today.

There is no evolutionary puzzle as to why the concentrated sources of essential energy, sugar and fat, are irresistible, but why is consuming them so bad for us? Meals and drinks that are high in carbs and fat, combined with sedentary lifestyles that utilize very little of the energy consumed, are the main fac-tors responsible for a global pandemic of obesity. In the United States, one-third of the adult population is obese, defined as hav-ing a body mass index (BMI) of 30 or more. BMI is the ratio of a person's weight measured in kilograms divided by the square of their height in meters. Another third of adult Americans are over-weight, defined as having a BMI between 25 and 30, so fully two-thirds of the population is consuming more calories than they can burn, causing their bodies to store the excess as fat.

The situation is similar in many other developed countries. Two-thirds of British men are overweight or obese, and the average across western Europe as a whole is 61 percent. Slightly smaller percentages of women than men are overweight, and slightly more women than men are obese in North America and western Europe. The situation is not as bad in Asia. More than a quarter of Japanese men and 18 percent of Japanese women are overweight, but the percentages that are obese are very small (3 5 percent) by comparison with Western countries.

There is great variation among developing countries, but being overweight is a problem in many of them. In Egypt, 71 percent of men and 80 percent of women are overweight or obese. In Mex-ico, the figures for men and women are 67 percent and 71 per-cent. These percentages are lower in other developing countries, but the combined population of the developing world is large, and so 62 percent of the global total of obese people live in developing nations. Hunger has not gone away, but these statistics paint a startling contrast with the familiar images of poverty and under-nourishment that used to be associated with developing countries. India now suffers malnutrition of two kinds: a proportion of the population goes hungry, and a growing proportion eats too much.

Being overweight is a major risk factor for metabolic syndrome—a flock of diseases that cluster upon the obese like birds of ill omen: hypertension, cardiovascular disease, type 2 diabetes, and a bloodstream overburdened with triglycerides (fat) and the bad kind of cholesterol. Type 2 diabetes is a disorder of the system that the body uses to regulate its fuel supply—blood glucose. In the normal course of events, when you consume carbohydrates, there is a spike in the concentration of glucose in the blood, the pancreas responds by releasing the hormone insulin into your bloodstream, and this causes cells around the body to take up the glucose and ultimately to convert any excess into fat. This process forms a feedback loop that lowers blood sugar, causing insulin production to decline and things return to the fasting state. Type 2 diabetes is a chronic condition in which, over the course of many years, the cells of the body stop responding to insulin. As the disease progresses, the concentrations of insulin and glucose in the blood rise because the feedback loop that normally regulates them has broken down.

The rise in type 2 diabetes is a health problem with an evolutionary dimension because susceptibility to the disease runs in families, and this presents us with a puzzle. Since the disease impairs fertility in both men and women and shortens life by about eleven years, the genetic variants that cause susceptibility ought to have been removed from the population by natural selection long ago, and yet the high rates of the disease demonstrate that this did not happen. There are two possible explanations for this. The first is that until very recently the genes concerned were not deleterious because they only become harmful in people who are overweight. Before obesity became the problem it is today, few people carrying these genes were obese enough to become ill. According to this hypothesis, it is the combination of being overweight and having the genes for susceptibility to type 2 diabetes that is the problem, not the genes by themselves.

An alternative hypothesis is that the genes that confer susceptibility to type 2 diabetes today represent the hangover of a condition that once actually conferred an advantage. This idea was originally proposed in 1962 by James Neel, a medical scientist at the University of Michigan, who sought to explain why diabetes runs in families. He suggested that some people inherit a set of genes (called a genotype) that causes them to store more energy as fat than others eating the same diet. He proposed that this genotype would have been an advantage during Paleolithic times when food was intermittently available. He called this the "thrifty genotype," drawing an analogy with the benefits of being thrifty with money. Thrift provides savings for a rainy day, a thrifty genotype stores up fat for a hungry week. Neel's thrifty genotype hypothesis proposed that genes that were once favored by evolution when food was periodically scarce are now detrimental to their carriers in the modern environment where food is plentiful. In this modern situation, people with this genotype lay down too much fat and this leads to disease.

Nearly 60 years after it was proposed, the thrifty genotype hypothesis is still widely offered as an explanation of the modern diabetes epidemic. Great advances have been made in all relevant areas of science since 1962, so we can now ask how well it holds up against the evidence. To begin with, we should examine James Neel's premise that our physiology is Paleolithic, adapted to the vicissitudes of a hunter-gatherer existence that he and others believed must have been characterized by alternate bouts of feast and famine. This argument can be broken down into two parts: the first is the assumption that famines were common in Paleolithic times, and the second is that the advantage of being fat in famine years would outweigh the hazards of being fat the rest of the time. Both of these ideas have been challenged.

Evidence is of two kinds: from the lifestyle of modern hunter-gatherer societies such as the San Bushmen of southern Africa, whom it is presumed live today approximately as our Paleolithic ancestors once lived, and from the genetics of obesity. A recent study that compared the frequency of famines in societies practicing different modes of subsistence found that in fact hunter-gatherer societies suffer less famine than agriculturalists living in similar environments. Agriculture is a high-risk, high-gain mode of subsistence because populations can become very large in the good years and then be exposed to severe shortages when crops fail. Hunter-gatherers are less vulnerable to famine because they have smaller populations and a broader range of food types to draw upon. Furthermore, estimates of the BMI of living hunter-gatherers show that they are invariably at the skinny end of normal (BMI around 20) and show no propensity to put on fat that might see them through lean times, until they adopt modern lifestyles and diets.

Thus, as with the Paleo diet (chapter 1), yet another story we have been telling ourselves about how our Stone Age ancestors lived turns out to be more *Flintstones* than fact. However, this need not necessarily spell the end of the thrifty genotype hypothesis altogether if we allow a modification to the original idea. One could argue that if agriculture introduced famine to the human species, then the advantage of the thrifty genotype should be seen in these societies, not in earlier times. Modifying Neel's hypothesis in this way, could the thrifty genotype have evolved since the advent of agriculture, as this is the more famine-prone mode of subsistence?

This modified version of Neel's idea would require more rapid evolution and spread of the thrifty genotype because the Neolithic, when agriculture began, takes us back only 12,000 years at the most. However, other changes have occurred on that time-scale. Large-scale studies of human genomes have identified a number of genetic changes that have taken place in the last 12,000 years, but none of them show the spread of a genotype that predisposes people to type 2 diabetes. In fact, quite the reverse—there is genetic evidence that natural selection operating since the Neolithic has favored alleles (gene variants) that lower, not raise, the risk of type 2 diabetes in some populations.

Neither, perhaps, should we expect to find such a spread because the marginal predisposition to type 2 diabetes that James Neel was attempting to explain in 1962 has now become a global pandemic. While in 1962 it made sense to ask why some families are more prone to this disease than others, it makes much less sense to ask this question now that a substantial proportion of the entire human population is at risk through being overweight. In fact, it can be argued that we should no longer be looking for genes that predispose some people to type 2 diabetes, but for genes that prevent this from occurring in some lucky individuals.

Does the end of the thrifty genotype hypothesis mean that evolutionary biology cannot illuminate the causes of the diabetes pandemic? No, it does not, although the nature of the illumination it can offer shines upon a different facet of the problem. The relevant question is no longer why are some people more susceptible to this disease, but rather what is it about how human physiology has evolved that makes most of us so vulnerable? The corollary to this question is what has changed in the human diet that has revealed this global vulnerability in just a few decades? According to Dr. Robert Lustig, an endocrinologist at the University of California, San Francisco, there is an answer to both these questions that can be summed up in just one word: fructose.

Fructose is the sweeter, deadlier twin of the sugar glucose to which it is joined to make a molecule of sucrose. Gram for gram, fructose is twice as sweet as glucose, and it is the sugar that many plants dose their fruit with, making them super-attractive to animals including us. As fruits ripen, they become sweeter and more aromatic, attracting animals that will carry them away. In doing so, they transport the plant's seeds to a favorable location where they will be deposited in a growth-stimulating pile of dung. A fruit, then, is a disposable wrapper for the precious cargo of a plant's genes. The nutrients in a fruit are the taxi fare; the bird, the bat, or the primate that takes the fare is the transport; and, from a plant's point of view, the destination is a secure place in future generations.

Food and drink manufacturers practice the same ploy as fruit, using an enzyme to convert some of the glucose in corn syrup to fructose, yielding high-fructose corn syrup (HFCS). HFCS is so cheap, so sweet, and so darned yummy that manufacturers use it in many processed foods and most sodas. Fructose consumption has doubled in the last 30 years, and there is a growing volume of evidence that it has a primary role in obesity and the etiology of metabolic syndrome.

The conventional idea about diet and weight gain or loss is that the body is like a bank account for calories. In Charles Dickens's novel *David Copperfield*, Mr. Micawber advises David: "Annual income twenty pounds, annual expenditure nineteen nineteen six, result happiness. Annual income twenty pounds, annual expenditure twenty pounds ought and six, result misery." For misery read "starvation" and for happiness read "fatness," and the parallel between money and calories sounds almost exact. Actually, both the metaphor and its model are equally plausible, equally widely believed, and equally fallacious.

The reasons both are fallacious are similar, too. Money does not just flow in and out of compartments in the economy; its availability is regulated by a central bank that can store money, print (devalue) it, lend it, and so on. This is how national economies are run. Likewise, the body does not react passively to the balance between caloric intake and expenditure, but it regulates the whole process, including the rates at which calories are consumed, stored, and burned. Food intake is a complicated science that in its details can be influenced by a whole host of factors. Among those that psychologists have discovered influence us in restaurants, for example, are the design of the menu and the silverware, the names given to dishes, the color of the plate, the shape of a glass, background music, and the ambience of the room. And that is before we have even smelled or tasted the actual food!

Apart from such subtleties, there are three important hormones that regulate how much we eat and what happens to the calories it contains: ghrelin, which signals when the stomach is empty; insulin from the pancreas, which signals when blood glucose levels require reduction; and leptin, which is made by fat cells and signals that fat reserves have reached capacity. These three hormonal signals are all received in the brain by the hypothalamus, where a balancing act that regulates the energy economy of the body takes place. The problem with fructose is that, despite being a sugar and carrying the same calories as glucose, it is not perceived like a sugar by the body and it fails to trigger the regulatory hormones that limit energy consumption and storage.

Let's follow the 12 grams of sugar in a typical glass of orange juice to see how differently glucose and fructose are metabolized. In the stomach, the sucrose from the drink is split into its 50:50 components, fructose and glucose. Glucose in the stomach is sensed as food and begins to inhibit the hunger hormone ghrelin, but fructose doesn't have this effect and so its calories get a free pass into your body without triggering any feedback that says: "Stop! I'm full." Next, the sugars enter the bloodstream and circulate around the body. Glucose is used as a fuel by all the organs of the body, but fructose can only be metabolized by the liver. Thus, the glucose is shared among all organs, but virtually all of the fructose—or in other words, half the calories you have ingested in the drink—end up in the liver. The liver also takes about 20 percent of the glucose load, which when added to the fructose means that this one overworked organ has to metabolize 60 percent of the calories in a typically sugary drink.

However, the damage that fructose does is not to be simply measured out in coffee spoons. The physiological effect of a spoonful of fructose is greater than the same quantity of glucose because it is invisible, not just to satiety sensors in the stomach, but to the other mechanisms that govern fuel economy too. Glucose in the blood stimulates insulin production by the pancreas, causing the organs of the body to use this sugar or to store it in fat. Fat cells produce the hormone leptin, and when this rises as a consequence of storing excess calories, the hypothalamus calls a halt to lunch. Fructose, however, does not trigger insulin, so these calories cause no knock-on rise in leptin and there is no signal to the hypothalamus to call a halt. Consequently, we carry on eating.

Though fructose wears a cloak of invisibility against the vigilance of the guardians of overeating, this is not its worst effect on the body. If it just made you fat, that would be bad enough, but it has even more insidious effects. A study of obese patients with metabolic syndrome in which the fructose in their diet was substituted by starchy food delivering the same number of calories found that they lost weight and that it took only nine days for their metabolic status to begin to improve. This shows that fructose has an impact on metabolic syndrome that cannot simply be accounted for by its calorific content. Something else is happening too. Robert Lustig calls fructose a toxin.

A toxin is a substance that interferes with vital metabolic processes and has life-threatening consequences. A characteristic of all toxins is that their effects are dosage-dependent, and this is true of the harmful effects of fructose too. Small amounts of fructose, released slowly into the bloodstream—for example, from eating whole fruit—are manageable by the liver. But large amounts ingested on a regular basis cause dangerous fat deposition in the liver and lead to the manifold ailments of metabolic syndrome and type 2 diabetes. Unfortunately, fruit that has gone through the juicer or smoothie maker behaves in the stomach like a very sugary drink, not like whole fruit, because the fiber that slows up the absorption of fructose in whole fruit does not do so when mechanically blitzed.

What, you may now be wondering, has this to do with evolution? I'm glad that you asked. The point is that we need to be careful not to resort to what Marlene Zuk has called "paleofantasy" when trying to explain our modern condition. Of course, we are limited by our evolutionary history, and this must explain at one level why fructose is toxic to humans, while hummingbirds cannot survive without the stuff. Returning to the thrifty genotype hypothesis, given what we now know about fructose and metabolic syndrome suggests that what Neel saw in certain diabetes patients was just the extreme cases in a pattern of vulnerability that has now been exposed in nearly all of us. Evolution is not destiny; it is potential. Many foods demonstrate just this property, and none do so better than cheese.

# 11

## *Cheese—Dairying*

Milk is the only food that we can truthfully say has evolved expressly for our consumption. Cheese is the result of sharing this gift of evolution with other organisms that trade us a small fraction of its energy content in return for an inexhaustible bounty of flavors. Mammary glands and their sometimes prodigious output are so important to the sustenance and survival of all infant mammals that one might wonder how the ancestors of mammals got along without it. This is the kind of question that can be asked about any adaptation. Charles Darwin insisted in *On the Origin of Species* that evolution by natural selection is a gradual process and that nature does not make jumps, but rather takes tiny steps that cumulatively build to major change over very long periods of time. In fact, he regarded gradualism as so fundamental to evolution by natural selection that he made it a test of the theory and wrote: "If it could be demonstrated that any complex organ existed, which could not possibly have been formed by numerous, successive, slight modifications, my theory would absolutely break down."

Zoologist St. George Mivart (1827–1900)—perhaps feeling impelled to live up to his mythical dragon-slaying namesake—persistently attacked Darwin's theory along these lines, arguing that an incipient mammary gland in some long-ago ancestor of mammals would have been so rudimentary as to be useless to an infant: "Is it conceivable that the young of any animal was ever saved from destruction by accidentally sucking a drop of scarcely nutritious fluid from an accidentally hypertrophied cutaneous gland of its mother?" A loaded question if ever there was one.

As a young man, Mivart began as a supporter of Darwin's theory, but although he continued to be an evolutionist of a sort, his religious convictions later led him to challenge the universality of natural selection and the absence from the theory of any form of design or direction from God. By the time that Darwin came to write the sixth and last edition of *On the Origin of Species* in 1872, he found the need to devote the majority of an entirely new chapter to answering Mivart's various criticisms. In it Darwin wrote: "The mammary glands are common to the whole class of mammals, and are indispensable for their existence; they must therefore have been developed at an extremely remote period. . . ." But Darwin goes on to say that Mivart does not argue fairly when he questions the value to offspring of a rudimentary mammary gland because such an organ was already known in the duck-billed platypus, whose young suck milk directly from ducts in their mother's skin. The skin is devoid of teats, and thus an infant platypus does what Mivart said was so improbable.

The platypus belongs to a strange group of egg-laying mammals called the monotremes, which are thought to resemble early mammals. The platypus is found only in the wilds of Australia and is nocturnal, hiding in deep burrows during the day. In 1872 the idea that they laid eggs was only an unconfirmed rumor. If Darwin had known for sure that the platypus not only has rudimentary mammary glands but also lays eggs, he could certainly have argued with even greater force that this animal is a relict, representing a transitionary phase between mammals' egg-laying ancestors and their later teat-equipped mammary perfection.

As Darwin suspected, the glands that produce milk are anatomically similar to sweat glands found at the base of hairs on the skin, and they almost certainly evolved as specialized versions of them. He was also right about the ancient origins of lactation, which genetic and biochemical evidence now suggest stretch back far before the first mammals appeared around 200 million years ago. The evidence for this is that all species, including the platypus, produce milk that contains the same basic array of constituents made by the same genes. This could only happen if all mammals descended from a common ancestor that had the full lactation kit to start with. Since this complex kit had to itself have time to evolve, lactation must have originated much more than 200 million years ago. It sounds paradoxical, but just as bird eggs are pre-avian, mammaries and milk are pre-mammalian.

Mammal milk is a unique fluid with two interdependent functions: to nourish infants and to protect them. The nourishment comes from the protein, fat, sugar (lactose), calcium, and other minerals in milk. The protection is provided by various antibodies and enzymes that have anti-bacterial action. These are especially abundant in colostrum, which is the first milk that a newborn baby mammal receives and which also contains immune cells from the mother.

It is remarkable, not to say odd, that the carbohydrate content of milk is entirely in the form of the unusual sugar lactose, rather than the universal sugar glucose that can be used by all cells. Why do mammals feed their young a carbohydrate that must be digested before it can be used? Surely, if milk had the immediate pick-me-up properties of a high-energy glucose drink, wouldn't that be better for baby? The answer may be that the very uniqueness of lactose provides an advantage over glucose. The world is full of bacteria and yeast that are hungry for glucose, while only a few kinds of bacteria are capable of utilizing lactose. Imagine the havoc that a bacterial or yeast infection of the mammary gland could cause mother and infant. In fact, brewers take advantage of the inability of yeast to ferment lactose by adding this sugar to sweeten a beer called milk stout. If glucose or sucrose were used instead, the yeast would turn it into alcohol.

There is a problem with feeding baby an unusual sugar, which is that the infant needs an unusual enzyme to break it down into a usable form. The stomachs of infant mammals are equipped with an enzyme called lactase that does just this. As the infant grows and is weaned, the production of lactase declines and then ceases altogether because it is not required. Lactose is not present in the foods that adults eat. As a consequence, adult mammals are normally unable to digest lactose, even though they were raised on it with their mother's milk. Inability to digest lactose is the normal condition of adult humans too. If you are lactose intolerant and consume unfermented fresh milk, this will cause diarrhea and stomach cramps as bacteria in your gut gorge themselves on lactose and inflate your guts with gas. If you are lactose tolerant, then you are the carrier of an allele that causes the persistence of lactase production into adulthood, and the story of how such mutations arose and spread belongs to your personal family history.

The first farmers who domesticated cattle and sheep around 11,000 years ago in southwest Asia (chapter 7) probably milked their animals as well as exploiting them for meat. However, adults would not have drunk milk because the first farmers, like their descendants in southwest Asia today, would have been lactose intolerant. Instead, they would have used milk to make yogurt, just as people in the region still do. Yogurt is made by mixing lactic acid bacteria (LAB) from a starter culture into the milk. Lactic acid bacteria possess the rare ability, missing from most bacteria and even lacking in our own cells, of being able to use lactose as an energy source. LAB feed on lactose, producing lactic acid as a waste product of their growth. Because the lactobacilli that ferment milk into yogurt use up the lactose, the product is safe for someone who is lactose intolerant to consume.

Yogurt production exploits properties of milk that evolved as part of its function as baby food. The proteins in milk are of two types: caseins, which precipitate out into curds when milk is acidified, and whey proteins, which remain behind in solution. Individual casein molecules are tiny fibers that flock together into something like round, nano-sized furballs called micelles. While the casein micelles remain suspended in the milk, they scatter light and give it its white color, but when they are removed from suspension, the whey that remains behind becomes clear.

The transformation of casein from suspension in milk to solid curds when the milk is acidified has an adaptive function for both mother and infant. It ensures that the milk can flow freely through the mammary gland to the infant, and that when the milk reaches the more acidic conditions of the infant's stomach, the casein will precipitate. This is necessary because casein takes many hours to digest and could be lost if it remained in suspension. In contrast, whey proteins, which remain in solution, are easier and quicker to digest.

Cheese makers use starter cultures of LAB too, and consequently such cheeses are also lactose-free. Cheese makers also use an enzyme called rennet from the stomach of calves that reduces the solubility of casein micelles, helping them to precipitate. Rennet is also made by certain plants including thistles, which provide an alternative source of the enzyme.

Milk residues on potsherds found at archaeological sites show that by 7,000 years ago dairy products were being used all over southwest Asia, especially where cattle herding was practiced. We do not know exactly what the dairy product stored in the earliest of these pottery finds was, but it was probably yogurt rather than cheese, which seems to have been a later development. Cheesemaking equipment is not found till about 1,000 years after the appearance of the first dairy pottery. Then, about 6,000 years ago, a new kind of pot appeared, pierced with many small holes. Fragments of these pots have the residues of milk fat on them, suggesting that they were used as strainers to separate fat-rich curds from lactose-containing whey to produce curd cheese.

Like the people who invented dairying in southwest Asia, the earliest Neolithic farmers in Europe were also lactose intolerant as adults. However, a lactase persistence mutation arose in the Caucasus Mountains of central Europe about 7,500 years ago. This mutation, which makes its carriers lactose tolerant as adults, swept through northern Europe and shaped the evolutionary heritage of people of European descent, wherever they are living today. In Utah, for example, more than 90 percent of the adult population is lactose tolerant.

Why did the lactase persistence mutation spread so quickly in Europe, but fail to evolve and spread in the original homeland of dairying? The second part of this question is more easily answered than the first. This is because the technology of curd cheese and yogurt making, invented in southwest Asia, removes the lactose from milk, enabling people who are lactose intolerant to benefit from dairying without any ill effects. Thus, if a genetic mutation causing lactase persistence were to appear among people using milk in this way, it would not confer any evolutionary advantage on the individuals carrying it and therefore it would not spread. This must be why lactase persistence alleles are nearly as uncommon today among people living in southwest Asia, where cattle dairying began, as they are in the Far East, where there is no dairying tradition at all. Why dairying technology did not prevent the evolution of lactase persistence in Europe is the remaining question.

The rapid spread of lactase persistence from central Europe northward represents one of the strongest examples of positive natural selection known in humans. It has been estimated from the rate of spread of the lactase persistence allele that it must have conferred an advantage of up to 15 percent over the normal allele. This explains how it spread, but not why. Despite the evolutionary evidence that lactase persistence has such a strong advantage, it has proved surprisingly difficult to pin down exactly what the nutritional benefit of drinking fresh milk is. It has, for example, been suggested that it supplies essential vitamin D or calcium, or that it provides a famine food when crops fail, as they probably often did in northern Europe.

A problem with understanding many evolutionary events, including the very origin of life itself, is that they can be unique, and this makes it hard to separate cause from coincidence. There is no such problem in the case of lactase persistence because it has evolved several times. Lactase persistence is found in Saudi Arabia, where it is due to a different mutation than the European one, though both lie in the same gene. There are no dairy cattle in Saudi Arabia, but nomadic Bedouin drink camel milk, and the water content as well as the food value of milk is likely to have been important in the evolution of lactase persistence in the arid environment of the Arabian desert. Milk as a source of water may also have played a role in natural selection for lactase persistence in East Africa. In this region, there are three alleles that enable pastoralists in Tanzania, Kenya, and Sudan to drink the milk from their cattle. The three East African mutations are all independent of each other and of those in Europe and Saudi Arabia, making a total of at least five separate instances of the evolution of lactase persistence. Globally, about one-third of the human population is lactose tolerant. For the rest of us, it's hard cheese.

If milk is the most natural food a mammal can get, then by contrast cheese is probably the most artificial. Everything else that we eat, however highly bred it may be, has a close relative in nature. Cheese is different because it is not the product of one species, or even two, but a microcosm made of dozens of kinds of bacteria and fungi. Biologically speaking, a cheese is a microbiome—or a community of microbes. The nearest equivalent microbiome found in nature is in the soil, which is also packed with fungi, bacteria, and other microbes that feed on dead material as well as each other.

The development of fast, cheap DNA sequencing has made identifying the diversity of bacteria and fungi in microbiomes a whole lot easier, and the result is that scientists exploring the cheese microbiome now turn up new discoveries at a rate not seen since Victorian naturalists first entered the Amazon rainforest with a twelve-bore and a butterfly net. For example, one small survey of Irish cheeses found five genera of bacteria not seen in cheese before. In terms of biodiversity, this is the equivalent of sitting down at dinner and finding that in addition to *Homo*, you are sharing the cheese board with *Australopithecus*, a chimp, and a gorilla.

Some of the microbes in cheese, like the mold *Penicillium camemberti* present in soft cheeses, have never been found anywhere else and have evolved in this specialized habitat from ancestors that are naturally at home in soil, in dung, or on the skin of cheese makers. Others have even stranger evolutionary origins, such as the bacteria from marine environments found living in the rinds of many washed cheeses. These may have made the leap from ocean to dairy in the sea salt used by cheese makers to treat their products.

A commercially important lactic acid bacterium used to make mozzarella cheese and yogurt is *Streptococcus thermophilus*. This harmless bacterium evolved from a pathogenic ancestor that it shares with the nasty *Streptococcus* species that cause strep throat and pneumonia. It is safe to eat mozzarella and yogurt because in the process of adaptation to living in milk, the genes that make its relatives harmful were disabled by mutation in *S. thermophilus*.

When the cheese mold that rejoices in the name *Scopulariopsis brevicaulis* is not doing duty in the dairy, it has been found hanging out on skin, in soil, in wheat straw, and in the seeds that a kangaroo rat had stored in its cheek pouch. By contrast, its relative *Scopulariopsis candida* seems more closely wedded to cheese environments, though it has also been found in the pages of a book. Science does not record whether *S. candida* has a preference for fiction or nonfiction.

In contrast to the mold *Penicillium camemberti* found only in cheese, *Penicillium roqueforti*, which is responsible for the blue veins found in Roquefort cheese, is a hobo that turns up all over the place. This particular fungus has been discovered in silage (fermenting grass), brioche, stewed fruit, wood, strawberry sorbet, and living on the inner walls of refrigerators. Blue cheese is produced in many different countries, but all are made with a strain of *P. roqueforti*. A comparison of the genetics of this mold among samples taken from French Roquefort and Bleu d'Auvergne, Italian Gorgonzola, Danish Blue, and English Stilton found that each strain of the mold was recognizably different from the others, suggesting that it was independently domesticated from a wild source in each blue cheese region. If you want to set your dinner guests a riddle over the cheese course, ask them what blue cheese and pigs have in common. Apart from both being full of fat and delicious, *P. roqueforti* and pigs (chapter 7) have both been domesticated many times.

The first step in cheese making takes place when so-called starter lactic acid bacteria (SLAB) turn the lactose in milk into lactic acid, precipitating the curds by mimicking the effect of acid in an infant mammal's stomach. The raw milk typically used to produce artisanal cheeses by traditional methods contains hundreds of bacterial species, including SLAB, that kick the cheese-making process off naturally. SLAB have to be added to the pasteurized milk that is used in industrial cheese making. What happens next depends upon the bacteria and fungi present in the milk and how this microbial community develops.

Cheese makers have four main levers at their disposal that enable them to control the development of the cheese microbiome and thereby determine the final flavor produced. They can add certain microbes such as *Penicillium roqueforti* or *P. camemberti* directly, control the ambient temperature of the cheese, control its effective moisture content (often by salting), and vary the length of time in storage. With these basic environmental parameters set by the cheese maker, the rest of the work is done by the microbes. A major player among the SLAB is the bacterium *Lactococcus lactis*, which feeds on the casein protein in curds, breaking it down into more than a 100 different fragments and producing the characteristic flavor and aroma of cheese.

Astonishingly, despite its critical role in cheese making, *Lactococcus lactis* appears to have evolved from a wild strain that lacked the genes that are essential to any microbe that lives in milk. The ancestral wild-type *L. lactis* lived on plants and had neither the lactase gene required to feed on lactose nor the genes required to break down casein. We are accustomed to thinking of evolution as a slow process, but when the selection pressure is high and generation time is short, change can happen very quickly. Both of these conditions would have applied to the Neil Armstrong of the *Lactococcus lactis* species—the first of its kind to enter the new world that would henceforth define its place in nature. Despite rumors that his landing place was made of cheese, Neil Armstrong found otherwise. Not so, the pioneer bacterium that first ventured, perhaps by the light of the moon, into milk. How did the first *L. lactis* turn milk to cheese?

Rapid multiplication is important in microbial evolution, but so is another process that is very common among bacteria but much rarer in multicellular organisms—horizontal gene transfer. The process that we think of as normal inheritance occurs when genes are transmitted vertically, from parents to offspring. This accounts for the resemblance between the generations. Horizontal gene transfer is the transmission of DNA between individuals of the same generation. It is the equivalent of a person who is lactose intolerant getting onto a crowded bus, riding half a dozen stops in the company of people who are lactose tolerant, and then getting off the bus with the newly acquired ability to digest fresh milk and also able to pass this ability on to any subsequent children by vertical inheritance. This will not happen to you on a bus however many stops you ride, but you could catch a virus from a fellow passenger and this is not unlike what happens in horizontal gene transfer among bacteria.

Viruses are able to insert their hereditary material (DNA or RNA) into a cell and hijack its DNA-replicating machinery, causing it to make more virus. In a similar way, segments of DNA called plasmids are able to insert themselves into bacterial cells, conferring new genes and new abilities upon the recipient. This is how that paradoxical creature, the lactase-deficient milk bacterium *Lactococcus lactis*, acquired the genes needed to live in milk. The vessel in which *L. lactis* acquired the plasmids was probably a cow's gut, and the fellow passengers that provided these useful genes must have been other species of bacteria that were already genetically equipped to ferment milk, perhaps used when the animal was a suckling calf.

I know that the tale of how *Lactococcus lactis* got its milk genes sounds like a just-so story straight out of Rudyard Kipling's imagination, but, oh, dearly beloved, it is no fable because important parts of the process have been reproduced experimentally. Scientists isolated *L. lactis* from bean sprouts and mixed the bacteria with milk. Samples of the treated milk were then taken after a few hours and used to start a new culture in fresh milk. This procedure was repeated for a thousand bacterial generations, taking about five months. At the end of the experiment, *L. lactis* was fermenting lactose and degrading casein like a milk native, which is in fact what it had become.

Not only had the bacteria evolved the ability to live in milk, but they had also lost some of the genes required by their old way of life. They were no longer able to ferment the kinds of sugars found in plants, nor could they synthesize some of the amino acids that they previously made for themselves. The plant sugars had been replaced by lactose, and although the amino acids were still needed, they were now available from breaking down milk proteins, thus making the bacterial genes that were previously used to synthesize them redundant. How Darwin would have loved to read the results of this experiment, for in *The Origin of Species* he spent several pages discussing how the effects of disuse led to the loss of previously important functions. He would have been doubly interested in the underlying genetics of such changes because the subject as we understand it today was an unwritten book in 1859.

As the SLAB in a cheese culture use up all the lactose and change the chemistry of the cheese, other bacteria and fungi colonize and grow. The activity of these microbes produce more changes, adding further to flavor as the microbiome develops. For example, in Swiss cheeses such as Emmental, the lactic acid produced by LAB is fed upon by propionic acid bacteria (PAB), which give cheeses of this particular type their characteristically nutty flavor. The growth of LAB is inhibited by the accumulation of lactic acid, which is in effect a waste product. Hence, the presence in Swiss cheese of PAB that feed upon and remove lactic acid stimulates the growth of LAB. The mutually beneficial relationship between LAB and PAB is a mutualism.

Mutualisms are of great theoretical interest in evolutionary biology because they challenge the notion that natural selection favors only selfishness. How can individuals driven by selfish genes cooperate? In theory, relationships of this kind are prone to the evolution of cheats—individuals that will exploit the cooperative behavior of others by taking and not giving in exchange, causing mutualisms to break down, or even preventing them from ever getting started. The LAB-PAB mutualism in Swiss cheese demonstrates one way out of this problem. In this case, the mutualism is stable because PAB feed on a waste product of LAB, and therefore cheating is not possible.

Mutualistic relationships between other lactic acid bacteria are more complicated. The two bacteria *Streptococcus thermophilus* and *Lactobacillus bulgaricus* cooperate in the fermentation of yogurt, each producing substances that stimulate the growth of the other. In *S. thermophilus* the genes for breaking down milk proteins have been lost, and these bacteria depend upon amino acids and peptides released by *L. bulgaricus*. For its part, *L. bulgaricus* uses a number of organic acids only produced by *S. thermophilus*. Unlike in the LAB-PAB mutualism, the substances traded between the different bacteria are not waste products, so how does a cooperative relationship like this get started?

The key to understanding the evolution of cooperation among bacteria is that these one-celled organisms are leaky, so that some quantity of essential resources that are of benefit to other cells are unavoidably released into the environment. Thus, it is quite possible for an individual bacterium to scavenge essential resources made by others and in so doing to save the energy cost of manufacturing the resource itself. The saved energy can be directed to another function that will increase its reproductive rate, thus giving the bacterium an advantage.

In such a situation, a mutation that, for example, knocks out a gene required for breaking down protein will benefit the carriers of the new mutation even though the function being disabled was previously a vital one. The missing function is now supplied by another bacterium, and the resources saved by the mutant are put to good use. Now imagine that the same process occurs in another species of bacterium, but affecting the manufacture of a different essential molecule, and we will now have a pair of bacteria that have evolved mutual dependence. This has happened entirely through a process of natural selection operating on characteristics that increase the reproductive rate of the individual bacteria. Cooperation of this kind arises from mutual advantage, not self-sacrifice.

Cheese contains not only cooperators, but also competitors. *Lactococcus lactis* and other LAB produce small proteins called bacteriocins that are toxic to other bacteria, but to which they themselves are immune. Bacteriocins have evolved as weapons of war among the bacteria, but they incidentally have an important function in cheese production because they prevent food-spoilage bacteria from colonizing the product. This helps stabilize the bacterial composition of the cheese community. Indeed, a food additive called nisin that is added to processed cheese to prevent spoilage is a bacteriocin originally isolated from lactic acid bacteria.

Cheese-making fungi also possess chemical weapons against other members of the cheese microbiome. The two *Penicillium* cheese fungi, *P. roqueforti* in blue cheese and *P. camemberti* in soft cheese, share an identical stretch of DNA that includes genes that produce a toxin that kills yeast, another that produces an antifungal protein, and a third that is thought to be anti-microbial. This bundle of genes is absent from strains of *P. roqueforti* that do not live in cheese, indicating that it has been acquired repeatedly by horizontal gene transfer from another fungus belonging to the cheese microbiome, though which is not yet known.

It is remarkable that despite the complexity of the cheese microbiome, so long as the initial ingredients are right, the four controls exercised by cheese makers are sufficient to enable the same cheese type to be reproduced over and over again. Cheese-making traditions have manufactured microbiomes that, though they are not found in nature, are as stable as any that are. The bacteriocins, antibiotics, and mutualisms that have evolved in the cheese microbiome contribute to this reproducibility by stabilizing its composition.

Milk is our most natural food, and yet cheese is paradoxically the most artificial, having no equivalent in nature. The term "artificial" is used pejoratively when applied to food, but cheese demonstrates that there is nothing to be afraid of in delicious artifice. The food that is entirely free of artifice anywhere in its evolutionary acquaintance with our species is probably inedible. Milk and cheese illustrate perfectly the interdependence of our own evolution with that of the species we eat. The rapid evolution of lactase persistence in Europe, East Africa, and Saudi Arabia followed the domestication of cattle and camels. Starting only 6,000 years ago, new bacteria and new microbiomes were summoned forth from the microbial world by dairying. We are also in debt to the microbial world for another product of fermentation, but this one has much deeper evolutionary roots than cheese. It is time to pop some corks, naturally.

# 12

## *Wine and Beer— Intoxication*

The affinity between humans, alcohol, and yeast lies deeper than a skinful. The tiny alcohol molecule called ethanol that yeast conjures forth from grape and grain possesses the giant transformative power of a mind-altering drug. It is capable of elevating or depressing mood, inspiring or befuddling the wits, inflaming lust while quenching performance, and inducing aggression as well as slumber. What object of desire as contrary, quixotic, and beguiling as this could fail to fascinate, madden, and enslave a lover?

Alcohol has us bound in knots with sinews that run deep. The reason that it has such a hold upon us, for good as well as ill, is that ethanol is a toxin for which we have evolved a tolerance. Alcohol is different in this respect from other mind-altering drugs. Opium, cannabis, and cocaine exert their effects upon the brain by mimicking natural substances in the nervous system. The plants that produce opiates, cannabinoids, and cocaine have evolved these psychoactive compounds as weapons in their arms race with herbivores. These substances only happen to affect us because brain chemistry is similar across the animal kingdom. A heroin addict is a bystander casualty of the war between poppies and caterpillars.

By contrast, ethanol is a toxin with no functional equivalent in human metabolism. This is also true of poisons like strychnine or arsenic, but if ethanol were purely a toxin, then wine, beer, and spirits would be obscure concoctions confined to the poisons cabinet of the pharmacy. The difference between ethanol and other poisons is that we have been exposed to ethanol in our food for a very long time because the signature dish of the great apes is fruit cocktail. Fruit is the mainstay of the chimpanzee diet and must also have been important in the diet of the common ancestors that we shared with them more than 5 million years ago. Where there is ripe fruit, there is yeast; and where there is yeast, there will be alcohol. We are grape apes as much as great apes.

The bloom upon a ripening grape contains a film of microbes that have the fruit surrounded like a besieging army encamped around a fortress piled with provisions. When the berries are harvested and crushed in readiness for fermentation, the must contains hundreds of kinds of fungi and bacteria. Just as in the aging cheese microbiome (chapter 11), when grape must ferments, the microbial species present wax and wane, vying with one another for the kaleidoscopic smörgåsbord of carbohydrates and poisoning each other with waste products. Foremost among the comestibles is sugar, and chief among the poisonous effluvia is ethanol.

Even when it has not been deliberately added by the winemaker, the microbe that usually comes out on top in the ferment of the grape microbiome is the yeast that beats all-comers in consuming sugar and producing alcohol: brewer's yeast (*Saccharomyces cervisiae*). A dozen B-list yeasts with wonderful names like *Dekkera*, *Pichia*, and *Kloeckera* have supporting roles in vinification, but most are ultimately defeated by the rising concentration of alcohol that *Saccharomyces cerevisiae* produces and that only this species can tolerate.

Because of the central role of *Saccharomyces cerevisiae*, the evolutionary history of fermented drinks begins not 10,000 years ago with the domestication of the grape or grain, nor 200,000 years ago with the advent of modern humans, or even 10 million years ago with the diversification of the great apes, but 125–150 Ma in the Cretaceous period, when flowering plants bearing fruit emerged. This is when the ancestors of modern brewer's yeast began their career consuming sugar in fruit and evolved the ability to turn it into ethanol in the presence of air, laying the genetic foundations of an epoch-making drinks business and a profitable sideline in bakery.

The dedication of brewer's yeast to turning sugar into ethanol is a peculiar thing. Why not use the sugar directly in growth, as other yeasts do, rather than waste energy in producing ethanol? The answer, already hinted at, would appear to be that ethanol is a weapon that prevents competing species of yeast and bacteria from consuming the sugar. *S. cerevisiae* manufactures ethanol using an enzyme called alcohol dehydrogenase (ADH). The *ADH* gene coding for the enzyme was duplicated around 80 Ma, producing the two *ADH* genes that are found in modern *S. cerevisiae*. The two ADH enzymes differ from each other by only a couple of dozen amino acids in the 348 that comprise the whole protein, but they perform opposing tasks. One enzyme, coded for by the *ADH1* gene, carries on the original function that began with the evolution of the first fruit, producing ethanol. The enzyme encoded by the second gene (*ADH2*) converts ethanol into acetaldehyde, which is used in yeast metabolism.

The duplication of the original *ADH* gene represented a significant evolutionary advance. During the one-gene period, the strategy of *S. cerevisiae* might be called "beggar my neighbor" because ethanol was used to starve and poison competitors at the cost of sacrificing some of its own sugar supply that was irretrievably converted into ethanol. The evolution of the second gene and its ability to turn ethanol made by the first gene into acetaldehyde introduced a new strategy that has been called "make-accumulate-consume." Now, ethanol is a weapon and a food store, all rolled into one. The reaction that turns ethanol into acetaldehyde requires oxygen, which is why if you want to prevent yeast from using ADH2 to consume the ethanol, thereby undoing all the useful work of ADH1, you must keep air out of the fermentation vessel.

Making wine is merely the domesticated version of the natural fermentation that takes place when fruit rots. Fruit, with an incidental lacing of alcohol, was a staple foodstuff of our primate ancestors, and in this diet lies the origin of our tolerance for ethanol and hence our interest in manufacturing it. This hypothesis was, until recently, mere speculation but genetic evidence has now brought it near to fruition. The source of our alcohol tolerance is a human version of that old friend the enzyme alcohol dehydrogenase, or ADH4 to be precise. ADH4 metabolizes ethanol when it reaches a high concentration in the liver. Reconstruction of the evolution of the gene that codes for ADH4 has revealed that this inestimable friend of all bibulous primates mutated to its current form somewhere between 13 and 21 Ma. This is around the time that the last common ancestor of orangutans and humans lived. Never offer an orangutan a beer. He won't thank you for it. Our closer relative, the gorilla, on the other hand, carries the same *ADH4* mutation that we do, though whether you really want to go drinking with one I leave to your sober judgement.

The *ADH4* mutation changed just one amino acid in the protein sequence of the enzyme, which is approximately equivalent to changing just a single word in a couple of pages of this book, but the effect was to increase by 40 times its ability to break down alcohol. This change could have been advantageous in two distinct ways. First, it occurred at a time in our evolutionary history when the climate of our African homeland became drier and primates were adapting to a landscape with fewer trees and more savanna grassland, so they probably spent more time on the ground. Fruit scavenged from the ground is likely to be more rotten and contain more ethanol than fruit gathered from the tree, making the *ADH4* mutation a handy innovation to have. Personally, I like the idea that we traded hanging around in trees for hanging around in bars, but the evidence that becoming less arboreal was important in *ADH4* evolution is purely circumstantial and hard to test.

The second advantage that the *ADH4* mutation may have provided is much easier to evaluate than the first. It is that possessing an efficient alcohol-detoxifying enzyme would not only have increased the food supply because rotting fruit could now be safely eaten, but also because alcohol is an energy-rich food in its own right. Ethanol provides nearly twice the calories contained in the same quantity of carbohydrate, so all those jokes about a liquid lunch have a sound basis. Medically, of course, the joke has a darker side, and it has been argued by biologist Robert Dudley that the human taste for alcohol and the roots of alcoholism lay in our fruit-eating ancestry. The evolutionary history of *ADH4* certainly supports the idea that we are adapted to tolerate alcohol, but this does not explain why some people become addicted to it while others do not.

There is significant variation among people and between cultures in the use of alcohol and susceptibility to alcohol abuse. Some of this variation has a genetic basis in another alcohol dehydrogenase enzyme called ADH1B. A mutation in the gene coding for ADH1B is found at high frequency in China and Japan, where 75 percent of the population possess at least one copy of the allele called *ADH1B*2*. The same allele is carried by a fifth of people in southwest Asia but is rare among Europeans and Africans. Individuals possessing *ADH1B*2* are much less likely than other people to drink a lot or to be alcoholics. This is at first sight paradoxical because the mutation increases the rate at which ethanol is metabolized a hundred-fold. Recall that the *ADH4* mutation that increased the activity of the ADH4 enzyme early in the evolution of the great apes made us alcohol tolerant and therefore prone to drink more rather than less. Both ADH enzymes convert ethanol to acetaldehyde, which causes nausea and headaches and is responsible for hangovers. So how can mutations in the two genes that both increase ADH efficiency have such different outcomes?

The explanation is in the different alcohol concentrations at which the two enzymes operate. ADH4 works at high concentrations of alcohol, while ADH1B works at low concentrations. Because ADH1B works at low ethanol concentration, the greater efficiency of the enzyme produced by the *ADH1B*2* allele causes a rapid increase in acetaldehyde after the first wee tipple and this is strongly aversive. As a result, people with this allele rarely drink to excess. They are also significantly less likely to suffer from cardiovascular disease or certain types of stroke. By contrast, people who drink more heavily because they either ignore the effects of having *ADH1B*2* or because they do not have this allele can build up their tolerance of alcohol through habitual use thanks to ADH4, but if they do so it is at a cost to their health.

Imagine the liver to be like a tank that processes alcohol into acetaldehyde. This is a poisonous substance, so the quantity in the tank needs watching. The amount of acetaldehyde that accumulates in the tank is controlled by three things: (1) how much alcohol arrives via the bloodstream, (2) how quickly the alcohol is transformed into acetaldehyde by ADH enzymes, and (3) how quickly acetaldehyde is metabolized. This last step is performed in the liver by three enzymes called acetaldehyde dehydrogenases (ALDH). If the whole process operates smoothly, because you have imbibed in moderation or your enzymes are swiftly processing alcohol and acetaldehyde, then none will escape back into the bloodstream and you will be spared a hangover—but not everyone is so lucky.

Some people carry mutations in ALDH-coding genes that decrease the activity of the enzymes that metabolize acetaldehyde. Two such mutations are known—one that is found in northern Europe and another in East Asia. The allele for the latter mutation can occur at frequencies of up to 40 percent. Individuals with this genetic makeup have poorly functioning ALDH enzymes, causing acetaldehyde to accumulate quickly when they drink alcohol. The downside of this is that drinking alcohol causes a practically instant hangover, but the upside is that this is so aversive that people with just one copy of this allele very rarely become alcoholics. People who have two copies, one inherited from each parent, have the worst hangovers and consequently complete protection against becoming alcohol-dependent.

So, if you are of East Asian origin, the chances are quite high that drinking alcohol will disagree with you. This is because you have a good chance of possessing the allele that raises the efficiency of ADH1B and the one that lowers the efficiency of ALDH, causing your body to quickly accumulate acetaldehyde if you drink alcohol. The first enzyme regulates the rate at which acetaldehyde is produced and the second affects the rate at which it is broken down, so if you have both alleles, I'll lay odds that you are a teetotaler. Why these alleles are more common in East Asia than elsewhere has yet to be investigated, but it may have nothing to do with alcohol consumption since ADH and ALDH are also involved in other aspects of metabolism.

There are foods that, when taken with alcohol, produce unpleasant effects because they influence the production or removal of acetaldehyde. The ink cap *Coprinopsis atramentaria* is a fungus that becomes toxic only when eaten in combination with alcohol. The fungus contains a compound called coprine that inactivates

ALDH, causing the effects of a severe hangover within minutes of taking a drink. Foods that contain alcohol dehydrogenase have the potential to interact with alcohol too. A soft cheese made with the lactic acid bacterium *Lactococcus chungangensis* contained significant amounts of ADH and ALDH. When the cheese was fed experimentally to mice that were also given ethanol, it lowered the alcohol concentration in their blood. This species of *Lactococcus* is not one normally used to make cheese, but if it were to catch on, it could alter wine and cheese parties forever, at least among mice.

The long evolutionary acquaintance among yeast, fruit, and primates must surely mean that when humans learned to cultivate plants, it was a foregone conclusion that fermented drinks would soon result. Indeed, it has often been suggested that the first cereals were not raised to make bread, but to brew beer. Beer is not only nutritious, but alcoholic fermentation also controls harmful bacteria that might be present in the water supply. The earliest direct archaeological evidence of fermented drinks so far discovered comes from residues of fermented rice, honey, and fruit found inside pottery jars from the early Neolithic village of Jiahu in Henan Province, northern China. This 9,000-year-old brew was made with either grapes or hawthorn fruit. Though it is the earliest example we know of, it cannot actually have been the first, even in China.

There are many wild grapes native to China, but just one, *Vitis vinifera*, in Europe, and this species is the progenitor of the domesticated grape that has been used in winemaking for millennia. Wild grapes are rare but can still be found from North Africa to the Rhine River. The cultivated plant differs from its wild ancestor in a number of important ways. In the wild, male and female flowers occur on separate individuals, so that only half of all wild plants produce grapes and males are required for fertilization and fruit set. Domesticated vines have hermaphrodite flowers that contain both male and female sex organs, so every vine produces grapes. Cultivated grapes are larger, have higher sugar content, and occur in bigger bunches with better fruit set than is found in the wild.

The earliest archaeological evidence of wine production from *Vitis vinifera* comes from a Neolithic village in the Zagros Mountains of northern Iran, where a jar dated to 7000 BP was discovered that contained residues of grapes and of a tree resin that is traditionally added to wine to inhibit vinegar bacteria. Myrrh was used for the same purpose in antiquity, and modern Greek retsina owes its characteristic taste to tree resin. About 1,000 kilometers to the north of this site, in a cave in the Caucasus Mountains near the village of Areni in Armenia, the exciting finding of a 6,000-year-old wine-pressing floor has been made. This is constructed with a sloping floor smoothed with packed clay that descends into the neck of a sunken jar. Around the floor were discovered other large jars suitable for fermenting and storing wine and the desiccated remains of grapes, grape skins, and stalks. One of these jars and the collection vessel in the floor contained chemical residues of red grapes. Short of finding an ancient wine label reading "Red Wine 4000 BCE," the evidence points as clearly as could be to Areni being the site of the world's first known winery. Wine is still produced in Areni; sampling a bottle of this production purchased in New York City, Ian Tattersall and Rob DeSalle described it in their book *A Natural History of Wine* as "all bright red fruit and black cherries, with just enough texture to leave a lingering memory that made us eager for more."

Nikolai Vavilov (chapter 4) proposed that the grape vine was originally domesticated in the Caucasus, where wild vines grow prolifically to this day. We might hope that today genetics could be used to date and pinpoint the location of the first grape domestication, but this has proved difficult because since domestication, wild plants have continued to fertilize domesticated ones throughout their range, blurring any telltale genetic signature of the founding event. On the positive side, the trickle of genes from wild plants to cultivated ones has helped to maintain the genetic diversity of grape vines in the face of the countervailing force of viticulture, which prefers the uniformity that comes from propagating vines by cuttings.

Despite the limitations, genetics does corroborate the archaeological evidence that vine domestication took place in the last 10,000 years in the Caucasus and spread south from there into the Fertile Crescent, reaching Egypt 5,000 years ago and traveling westward around the Mediterranean into southern Europe, arriving in France about 2,500 years ago. According to some students of the vine, the thicket that is grape genetics contains evidence hidden within it that *Vitis vinifera* was independently domesticated in the western Mediterranean as well as in the Caucasus, but at the time of writing the jury is still out on this.

Such is the mystique of wine and the chauvinism of *terroir* that claims have been made for secondary domestications of the grape in Sardinia, Languedoc, and Spain, though nothing yet matches the justified pride of the Georgians in the primordial vines growing in their Caucasian mountain fastness. These are the progenitors toward which all tendrils must incline in homage. In truth, every vine, like every person, is the product of genes transmitted through history, genes adapted to place and the shaping of the individual through circumstance. With an estimated 10,000 cultivated grape varieties, *Vitis vinifera* is a rather good parallel for the diversity of our own species, though we lack one evolutionary trick that vines make very good use of: clonal growth.

Since Roman times, vines have been propagated by grafting cuttings of the chosen grape variety onto established rootstocks. This means that all the grapes belonging to a particular cultivar are the fruit of a single clone. New varieties have been produced by crossing existing ones, selecting among the progeny and then clonally propagating the new grape by grafting. Because the parents of new varieties were traditionally chosen from those already available locally, grape varieties are clustered in families of related clones. For example, a genetic analysis of the grape varieties growing in the north of Spain found that they are all closely related to one another. In post-Roman times, vines for winemaking were transported around Europe with the spread of Christianity to provide wine for use in Holy Communion. The original clone from which the Spanish varieties were derived appears to have been brought across the Pyrenees from France along a pilgrimage route to the Spanish city of Santiago de Compostela.

Clones do occasionally spontaneously produce mutant shoots called sports that are different from the rest of the plant. Grape varieties evolve by this means, subject to selection on the part of the grower, who chooses whether or not to propagate the sport.

Pinot is an old and "noble" grape, used in Champagne and Burgundy, that has evolved many new varieties by this means. Sixty-four such Pinot clones are recognized and registered in France alone. The great majority of mutations in grape clones are caused by pieces of maverick DNA called transposable elements. These are stretches of DNA sequence that can move about the genome and duplicate themselves, reaching great abundance. Transposable elements comprise 40 percent of the grape genome and half of our own.

Transposable elements are sometimes called "jumping genes," though they have no primary function such as the genes that code for proteins do. They are nonetheless of great evolutionary significance because of the mutations they cause when they insert themselves into functional genes, disabling or altering them. A common kind of mutation that has given rise to many new Pinot clones, such as Pinot Blanc (white) and Pinot Gris (gray), affects the color of the grape. The black color of Pinot Noir grapes, and indeed other black or red varieties, is due to the presence of a pigment called anthocyanin. Pinot Blanc, Pinot Gris, and other white grapes owe their lack of pigment to a mutation caused by a transposable element that disables a gene controlling the production of anthocyanin.

Transposable elements can jump out of functional genes as well into them, so reversing their effect. By this means, the red-skinned grapes Ruby Okuyama and Flame Muscat derive from the restoration of gene function in sports of the white grapes Italia and Muscat of Alexandria. The clones producing white grapes have two copies of the white (mutant) allele, but it turns out that most red and black varieties such as Pinot Noir, Syrah, and Merlot also have one copy of the white allele. This suggests that a single dose of anthocyanin is good, but that a double dose is too much. A possible explanation for this is that the production of pigment is costly to the plant and that natural selection, artificial selection, or both favor clones that have one copy of the anthocyanin-promoting gene turned off. This hypothesis is so far untested.

The practice of grafting, which is used to preserve the genetic identity (the genotype) of grape varieties, was to have another quite unexpected benefit when in the 1860s a new vine disease appeared in the South of France. Vine leaves dropped prematurely, grapes withered on the vine, and roots rotted. As is often the case when a new disease strikes, the cause was at first hard to fathom. Dead vines revealed no sign of what had killed them, and the culprit did not come to light until Jules Émile Planchon, professor of botany at the University of Montpellier, had the bright idea of examining the roots of still-healthy plants on the edge of affected areas. The roots of these plants were swarming with aphid-like bugs sucking at their sap.

The bug was hitherto unknown in Europe, and for nearly a decade Planchon strove to decipher the insect's complicated life cycle, which was eventually discovered to have no fewer than eighteen stages. Meanwhile, the disease continued to spread, devastating vineyards throughout France and appearing in Spain, Germany, and Italy. Soon the state entomologist of Missouri, Charles Riley, heard of the insect that was destroying the vineyards of Europe, and he wondered whether it might be the same one that had been reported living on the leaves of American vines in New York State. One puzzle was that while the American insect infested vine leaves, the one in France lived on vine roots. In 1871 Riley visited France, where he saw the bugs for himself, and it became clear that the insects in America and France were one and the same species. This is the insect now known as *Daktulosphaira vitifoliae*, or more pronounceably as phylloxera.

Riley was an early follower of Darwin, and he saw the phylloxera problem from an evolutionary angle. He reasoned that since the insect was native to America, vines there would be adapted to resist its damage. Hence the solution to Europe's phylloxera epidemic would be to graft European varieties of *Vitis vinifera* on imported, resistant rootstocks of American vines such as *Vitis riparia*. In 1873 Planchon toured the United States and saw American vineyards planted with *Vitis vinifera* by European immigrants who hoped in vain to reproduce the wines of their homelands. In contrast to these doomed plantations, Planchon was shown a thriving European grape grafted onto a wild grape from Texas. He heard that grafting onto native American rootstocks was the only way that *V. vinifera* could be grown in the United States. At first there was reluctance in France to believe that the country that was the source of the problem might also provide the solution, but eventually grafting onto phylloxera-resistant rootstocks saved the European wine industry and its ancient varieties. Riley's contribution was recognized by his appointment to the distinguished French Legion of Honor.

Riley's evolutionary insight also led him to caution that the American Concorde grape—which is a hybrid, not a graft, between an American species and the susceptible European *Vitis vinifera*— might allow phylloxera to evolve a form that is able to attack the American cultivar, even though it was free of the disease in his day. A century later Riley's prediction came true, and there is now a phylloxera that is specifically adapted to living on the Concorde grape in the United States.

The phylloxera crisis caused a loss of genetic diversity in European wine varieties since not every clone was saved by being grafted onto American rootstock. One variety that was thought to have been lost forever was a red grape called Carménère, but thanks to the popularity of French vines around the world in the nineteenth century, it was later discovered that this variety had survived in China and in Chile, where phylloxera is absent.

The effects of humans on the evolution of the ingredients of alcoholic drinks are by no means confined to how we have shaped the grape. Yeast also bears the hallmarks of our thirst in its genome. *Saccharomyces cerevisiae* has a worldwide distribution, but the strains used to make local drinks are independently domesticated from the wild in each region. The yeast strains used to make European wine have a common origin in the Mediterranean, though some of them show up in the United States doing the same job in wineries there. Sake in Japan is made with its own local breed of *S. cerevisiae*, as is rice wine in China, palm wine in Nigeria, and rum in Brazil. Each of these drinking yeasts was independently recruited to their task from local populations of wild *S. cerevisiae*, demonstrating through repetition just how well adapted this species is to the niche that we have created for it, wherever that may be.

In the course of working out where the domesticated yeast used in different parts of the world come from, samples have been collected from the wild as well as from vineyards and wineries, and the wild samples have revealed an unexpected side to the natural history of *S. cerevisiae*. Fruit is a highly seasonal resource in Mediterranean environments, so where yeast spends the rest of the year and how it finds its way around has always been a bit of a mystery. It has now been found living year-round on the bark of oak trees, where it probably feeds on sap. *S. cerevisiae* has also been found in the guts of the European hornet. Hornets feed on ripe grapes and are at their most abundant in vineyards at harvest time, providing an ecological link between wild and vineyard populations of yeast. The hornet-yeast reservoir is a perennial one that is passed from one generation to the next when adults feed larvae.

Nowadays, selected yeast strains are cultured and added to wine and beer, reducing the scope for gene exchange with wild populations. However, some winemakers and brewers do still use wild yeast. The Rogue craft brewery in Newport, Oregon, has taken the name "brewer's yeast" quite literally, creating a beer that is fermented with "wild" yeast cultured from the brewmaster's beard. Even in clean-shaven breweries, if such exist, yeast is continually present since beer making is not confined to the season of the barley harvest in the way that wine production is limited by the grape harvest. Barley and other grains used in brewing beer store very well and are ready to yield their sustenance whenever required because this is precisely what seeds are adapted to do in nature, providing plants' offspring with nutrition when germination takes place. Germinating grain is also the first step in brewing beer, activating the enzymes that are needed to release the sugar for yeast to work upon.

Natural and beverage environments require different adaptations and apply different sources of natural selection to yeast. In adapting to the wine environment, *S. cerevisiae* has acquired three separate, foreign stretches of DNA containing 39 genes by horizontal gene transfer from other species of yeast. These genes perform important wine-related functions like helping wine yeast to utilize the range of different sugars, amino acids, and nitrogen sources present in fermenting grape must. This phenomenon illustrates how fluid the genomes of microbes are, with the frequent horizontal movement of genes between species that are too distantly related to hybridize through sexual reproduction.

Adaptation to the wine environment has gone even further in specialized strains of *S. cervisiae* called flor yeast. These strains are part of the aging process in a number of white wines, particularly fortified ones like sherry that are matured in the cask. Flor yeast are uniquely adapted for growth at the end of fermentation when all the glucose and oxygen in the wine have been used up and the concentration of ethanol has reached a maximum. Normal yeast cannot grow in these conditions, but flor yeast are adapted to thrive in them. Low-glucose concentration in the wine activates the expression of a gene called *FLO11* that makes the surface of the flor yeast cell water-repellant. This causes flor yeast cells to stick to one another, trapping bubbles of $CO_2$ gas, which then floats them to the surface of the wine. At the surface, the cells form a bio-film called a flor, from which these strains get their name. Floating at the interface between the wine and the air in the top of the cask, flor yeast have the best of two worlds: access to the ethanol in the wine beneath and the oxygen in the air above, the combination of which enables them to use ethanol as an energy source.

*Saccharomyces cerevisiae* is not as good as other yeasts at fermentation at low temperature, but no other yeast is as alcoholic. The yeasts found in low-temperature fermentations turn out to be hybrids between *S. cerevisiae* and other species of *Saccharomyces* that function better at low temperature, combining the strengths of both. These hybrids have evolved several times independently in the fermentation of lager beer. The full story of how the lager yeast called *Saccharomyces carlsbergensis* came to inhabit the vats of the Copenhagen brewery where it was isolated and after which it is named has yet to be nailed down. The part of the story that has so far been deciphered from its genetic relationships has the elements of a Norse saga.

The genes tell us that *Saccharomyces carlsbergensis* is a hybrid between that ubiquitous drunkard *S. cerevisiae* and a cold-tolerant cousin called *Saccharomyces eubayanus*, but there are two possible sources for the latter parent. This cold-adapted species has been found living on the bark of Southern beech trees in Patagonia in the far south of Argentina and also in Asia, on the Tibetan plateau.

Strains of *S. eubayanus* from both these places are more than 99 percent similar to the parts of the genome of *S. carlsbergensis* that do not match its other parent, so it would require the judgement of Solomon and some more forensic evidence to decide which of the two frigid regions, the tip of South America or the top of Asia, supplied lager yeast with its cold-adapted genes. Whichever it was, the bigger mystery is how the genes got from either of these remote locations to Europe, but somehow they did.

In 1845 the founder of the Carlsberg brewery, Jacob Christian Jacobsen, obtained the yeast for his beer from the Spaten Brewery in Munich, Germany. Whether this German yeast inoculum contained *S. eubayanus* itself or its hybrid with *S. cerevisiae* we do not know, but it contained the raw genetic material of what was to become *S. carlsbergensis*. This yeast culture was used to make Carlsberg beer for 38 years. By a back-of-the-beer-mat calculation, we can reckon that if a new brew was made each week over that period, this represented about 2,000 serial cultures and tens of thousands of yeast generations of selection. What this means is that whatever went into the first fermentation, the result after 38 years was a yeast population that had been completely domesticated by Danish brewing. Then, in 1883 a microbiologist called Emil Christian Hansen working at the Carlsberg laboratory made a pure culture of *S. carlsbergensis*. This allowed control over the quality of the beer produced, changing the face of brewing and forming the foundation of the multinational operation that Carlsberg was to become.

The story of the Carlsberg yeast could so easily be retold in the guise of a Hans Christian Andersen fairy tale of how a drunken reprobate was made respectable by a wandering vagabond from a far-off land and how their talented and virtuous offspring went on to worldwide fame and fortune. Equally, the twists and turns in the genetic history of the grape could supply the plot of a 26-part television drama. Only the names of the parties involved need be changed in retelling it as a human drama, the saga of illicit unions reviving ancient and decrepit bloodlines, sex change, plagues, and the return of long-lost relatives. Like the path of true love, the course of evolution never does run straight.

Alcohol is best enjoyed in a social environment, for there it fosters bonhomie and liberates the free flow of conversation and wit that the Irish call the craic. We are all wittier on wine, or at least so we would like to think. Supply food as well as alcohol, both in liberal quantities, and a mere dinner shall become a feast. Feasting takes our exploration of evolution and food into the social domain.

# 13

## *Feasting—Society*

To share food is human, whether in feast or famine. The impetus to share is engraved in the human psyche, but this does not mean that all have equal access to food. Would that it were that simple. Rather, food is entangled in social relations with all the attendant complications that this brings. Nowhere is this more poignantly illustrated than in the history of Ethiopia, land of feast and famine, homeland of Lucy *Australopithecus afarensis*, and cradle of our genus, *Homo*.

In 1887 Queen Taytu Bitul, third wife of Emperor Menelik II of Ethiopia, arranged a feast to celebrate the consecration of the church that had just been built in the new capital, Addis Ababa. The feast was planned on an imperial scale befitting the military achievements of Menelik II, who, by defeating his neighbors including a colonial Italian army, had restored Ethiopia to something nearer the status enjoyed by the ancient kingdom founded nearly 2,000 years before. In a vast tent erected in the courtyard of the Entoto Maryam church, situated on a mountain top overlooking the city, Queen Taytu created a feast that would enter the history books.

In five days of feasting, the guests consumed *wet* (stews) containing the meat of more than 5,000 oxen, cows, sheep, and goats. To this day, Ethiopia has more livestock than any other African country. The royal favorites at the feast had special dishes prepared for them, including "minced beef sautéed in spiced red pepper, well done . . . mutton ribs and pieces in pepper stew . . . slightly rare beef in spicy sauce . . . mutton ribs in a turmeric-spiced broth . . . and peppered ground pea sauce cooked with bits of meat."

*Wet* is eaten with the fingers in a fold of injera, a large, spongy pancake made with the fermented flour of the native domesticated grain, teff. Injera was served from a thousand baskets that circulated among the diners, kept supplied by five separate kitchens. Forty-five massive clay jars of butter spiced with ground red pepper were also paraded among the diners, and the guests' thirst was slaked from an equal number of jars of tej, a honey wine. The drink was replenished from twelve pipes that were fed by gravity from an uphill reservoir of tej. Guests of lower rank received a smoky beer made from malted and roasted barley.

Taytu's feast of Maryam drew upon the cuisines of all her husband's conquered territories and was an instrument of statecraft designed to overwhelm their subjects and guests with culinary shock and awe. The official chronicler of the feast recorded that the combined aroma of tej and freshly made injera caused a faintness of the heart. However, this feast was not a sign of good things to come but a harbinger of famine. Over the ensuing five years, a combination of drought and the spread of rinderpest resulted in the loss of 90 percent of cattle and the death of a third of the human population.

Ethiopia's precarious history has been punctuated with recurring drought and famine since at least 250 BCE, but some of the worst famines have occurred in recent times when the country endured a perfect storm of drought, conflict, population pressure, environmental degradation, and totalitarian government. These factors all came to a head in the period 1983–85 when famine overtook 8 million people, killing between 600,000 and 1 million. Such was the scale of the famine that the normal impulse of people to help each other broke down. A third of Ethiopian households shared food and money with starving relatives, but the majority struggled even to feed themselves. People avoided their relatives rather than face the shame of being unable or unwilling to share.

Governments were slow to come to the aid of starving Ethiopians, but horrific television pictures of emaciated men, women, and children mobilized a huge response from the public around the world. By the end of 1984, individuals in Western countries had given more than $150 million for famine relief, the equivalent of nearly $450 million now.

This tragic episode demonstrates that when people can share food, or the means to buy it, they will do so even when the recipients are quite unrelated and indeed unknown to them. Sharing food is a paradigm example of altruism, defined as behavior that benefits someone else at a cost to the giver. A simple, arguably naive interpretation of natural selection would suggest that sharing resources with unknown people is not an adaptive behavior and that it therefore requires a particular evolutionary explanation. To some it may seem almost indecent to ask why we share; however, to ask how humans evolved to be sharing and caring is not to diminish these social traits, but to inquire how humans became humane.

Explaining the evolution of altruism has been a tough call for evolutionary theory since its beginning. In his book *The Descent of Man*, Darwin wrote of individuals possessing what he called "moral virtue" that "the circumstances, leading to an increase in the number of those thus endowed within the same tribe, are too complex to be clearly followed out." How can unselfishness evolve in a world governed by selfish genes? Three kinds of explanation have been proposed. The first is called kin selection and is based on the idea that our genes, including any hypothetical "gene for altruism," do not reside only in ourselves, but also exist as copies in our relatives. For example, normal siblings share half their genes, inherited from their parents.

The great twentieth-century evolutionary biologist and polymath J. B. S. Haldane said in one of the flashes of insightful wit for which he was famous, "I will lay down my life for eight cousins or two brothers." First cousins are more distantly related than siblings and share only one-eighth of their genes, so eight cousins are required to make the books of kin selection balance. In his personal behavior, Haldane was an extreme altruist who during the Second World War conducted dangerous experiments on himself to determine how submariners could safely be evacuated from stricken submarines. It is all too easy to imagine him sacrificing himself, even for non-kin.

Another British evolutionist, W. D. Hamilton, formalized this idea of kin selection by proving that for an inherited altruistic trait to spread, its cost to the giver must be less than the gain to the recipient multiplied by their degree of relatedness. In his quip, Haldane was implicitly assuming that the cost to him (as the giver) and the benefit to each of his cousins (the recipients) were equal. Each cousin has a degree of relatedness of one-eighth, hence the need for eight of them to match Haldane's self-sacrifice. In actuality, Haldane's example only brings into balance the costs and benefits of his altruism, while Hamilton's Rule says that the total of the benefits to kin must exceed the cost to the altruist. Nine cousins would do the trick.

Can kin selection explain why people share food? A complete answer to this question would require an experiment in which we would have to measure the costs and benefits of food sharing to see if they conform to Hamilton's Rule, but this is not at all easy. In the calculus of evolution, costs and benefits are measured in units of fitness, which represent the number of offspring contributed to future generations. Imagine trying to measure the fitness consequences for yourself and all of your guests of sharing your meals with relatives, versus the alternative of dining alone! In the absence of such an unlikely test of Hamilton's Rule, we have to resort to more circumstantial evidence.

Comparative studies of different human societies find that the most consistent pattern is preferential sharing with relatives, or as one review wryly put it: "Although agreeing on little else, anthropologists acknowledge that kinship is one of the central organizing features of human society." This pattern is consistent with what we would expect from kin selection, although anthropologists have not always accepted this. One objected that hunter-gatherers were not strong on arithmetic and so were unlikely to be able to direct their behavior according to Hamilton's Rule, much less could animals calculate their one-eighth relationship to their cousins. With characteristic acuity, Richard Dawkins remarked, "A snail shell is an exquisite logarithmic spiral, but where does the snail keep its log tables?"

In the case of Hamilton's Rule, this is not an instruction in some imaginary book of etiquette for hunter-gatherers, or even for their genes, but a way of keeping score in an evolutionary contest between individuals exhibiting different kinds of behavior toward others. If there is a genetically determined predisposition toward sharing with relatives, Hamilton's Rule tells us when this will be favored by kin selection. Kin selection is just a particular kind of natural selection. Let me add that kin selection does not provide a moral justification for nepotism, even if it does potentially explain it. Human society is far more sophisticated than that and often punishes unfair treatment.

Favoritism toward relatives is widespread in animal societies, although not all share food after weaning, even with their own offspring. Among primates, about half share food with offspring and about half of these also share among adults. Primates that do not share with offspring do not share with adults either. This suggests that in the evolution of sharing behavior, sharing with offspring is a precursor to sharing with adults, including non-kin. This is what would be expected from the gradual nature of evolutionary change: sharing begins with the nearest kin (offspring) and then develops with other adults, especially potential mates. Charity begins at home, as the proverb has it. So, the circumstantial evidence supports kin selection as a basis for the evolution of food sharing, but it cannot be the whole story since we do also share with non-kin. Why do we do that?

The second type of explanation seeks to explain the evolution of sharing among non-kin on the basis of reciprocity. Reciprocity might be direct, in that I share my food with you in the expectation that when I am hungry tomorrow you will share with me, or that we might have sex. The reciprocity could also be indirect. Indirect paybacks may take the less tangible form of friendship, mutual support, or kudos. Biologists have been in the habit of calling reciprocated acts "reciprocal altruism," though the phrase is falling out of use because it contains a contradiction in terms. An act is not altruistic in the proper sense of that word if it is performed in the expectation of future reward. I work in the expectation that I shall be paid at the end of the month, and neither I nor my employer regards ourselves as altruists. But, whatever you call it, does direct reciprocity explain food sharing?

If you ask people why they share with their friends, they will usually strenuously reject the idea that they do so in the expectation of gain. However, one could inquire further and ask whether they would become or remain friends with someone who never reciprocated, and I suspect the answer would be "no." The Roman orator Cicero (106–43 BCE), who lived in violent times when knowing whom you could trust was a matter of life and death, wrote: "There is no duty more indispensable than that of returning a kindness. All men distrust one forgetful of a benefit."

By chance, there survives a letter from the Roman author Pliny the Younger reproaching a friend who has broken the rules of amity by failing to appreciate an invitation to dinner. In it we learn what he missed at the feast:

> Dear Septicius Clarus: You promise, but you don't turn up to dinner, I'm afraid! All ready and waiting were a lettuce (each), three snails, two eggs, porridge . . . olives, beetroot, gourds, bulbs and a thousand things no less enviable. You could have listened to comic actors or a poetry reader or a lyrist, or, such is my generosity, all three. But you chose to go to someone else's, and what did you get? Oysters, sows' wombs, sea-urchins, and dancing girls from Cadiz!

As if to say, "How vulgar compared to the sophistication I had to offer."

Even if friendship is defined as a relationship of trust rather than a mere trading of favors, reciprocity is the foundation on which it is built. The comparative data on how hunter-gatherers share food illustrates this. In some tribes, people are expected to reciprocate and are ostracized if they never share back, while others seem to operate on a more indirect basis of what goes around comes around. In such cases, everyone is expected to share, without any strict accounting having been made. These are different way of organizing food sharing and social relations, but both are built on reciprocity, operating either directly or indirectly.

When the hypothesis of reciprocal altruism was first promulgated in the 1970s, there were believed to be many examples of this kind of behavior in animal societies. However, the motivations of animals are notoriously difficult to interpret without ambiguity, and on closer examination other explanations have been forthcoming for many of the cases. A good example of this difficulty is seen in how chimps at the Gombe National Park in Tanzania, famously first studied by Jane Goodall, hunt monkeys and share the spoils.

A hunt typically begins when a band of male chimps spots a monkey that is somewhat distant from the protection of its troop. One chimp gives chase, while other chimps in the band respond, not by forming a pack, but by spreading out to close off the monkey's possible escape routes or positioning themselves for an ambush. The earliest interpretations of such hunts described them as cooperative enterprises in which chimps played complementary roles, all in expectation of receiving a share of the meat at the end. The males in chimp bands are related to one another, so kin selection would be a possible explanation for how such cooperation had evolved.

With more field data available, another much more individualistic interpretation emerges. With one chimp driving the monkey, the best strategy for each of the remaining chimps is to position themselves in a possible path of escape or to hide and ambush the prey. Each chimp hopes to capture the prey himself because the chimp who kills the monkey gets most of the meat. What looks to us like a coordinated process because it resembles how we ourselves would hunt may in fact only be the outcome of each chimp behaving selfishly in his own best interests.

This individualistic interpretation of the monkey hunt is supported by how chimps behave with the food that they have captured. If direct reciprocity was involved, one would expect the meat to be willingly shared, but at Gombe the chimp who killed the monkey always tried to hold on to it and usually only shared it under duress. He would run away from the rest of the band, climbing inaccessible branches to where he might be able to dine unmolested. Usually, the other chimps would manage to gather round and try to take some of the prey, or cover the mouth of the captor with their hands to prevent him from eating. In these cir-

cumstances, a chimp with a monkey carcass had difficulty keeping it all to himself and allowed some meat to be taken so that he
would be left alone. This has been described as "tolerated theft."

Chimps at Gombe did sometimes share meat willingly with
non-kin, but in most cases it was not obvious why they did so with
some individuals rather than others. Quite possibly there were
friendships between such individuals that were unknown to the
human observers. Only the alpha male of the Gombe troop con-
sistently gave females meat he had captured, suggesting that this
food sharing was based on the existence of sexual relationships.

Chimp societies exhibit cultural differences, rather as human
societies do, and food sharing by both male and female chimps
has been observed in other places where it seems to play a greater
role in forging alliances than it does at Gombe. In one of these, at
Sonso in Uganda, there is remarkable evidence of how this behavior could have evolved. In mammals from voles to humans, the
hormone oxytocin lowers aggression and plays a role in social
bonding between mother and infant and between sexual part-
ners. A study of the wild chimpanzees at Sonso found that when
individuals shared food, the level of this hormone present in the
urine, which is a measure of how much is present in the bloodstream, increased in both the donor and the recipient. Thus food
sharing had the direct effect of increasing the social bond between
chimps, whether they were related to one another or not.

The response of oxytocin to sharing food shows not only how
this behavior can strengthen social bonds between kin and non-
kin alike, but reveals the mechanism by which food sharing among
adults could have evolved from food sharing by mothers with
their offspring for whom oxytocin is the glue that secures their
attachment. To be clear, the role of oxytocin does not tell us
why strengthening social bonds is favored by natural selection.
Rather, it shows how, in situations where it is of advantage such
as when mothers feed their offspring, physiology steers an animal
toward the kind of behavior that increases fitness. Oxytocin is the
servant, not the master of the genes. Our hormones propel us into
sexual behavior in just the same way.

This excursion into the food-sharing habits of chimpanzees
may appear to have taken us a long way from the crowded feast
of Queen Taytu or the lonesome dinner of Pliny the Younger, but

although chimps are our cousins and not our ancestors, they provide a reference point for the evolution of our own food habits. We are alike in that we both favor feeding our relatives as a bald consequence of kin selection. Chimps and humans are also hormonally primed to form social bonds by sharing food when it is advantageous to do so. However, beyond these biological essentials, the comparison between our two species highlights the evolutionary difference between us, rather than our similarity.

Tolerated theft of the kind seen among Gombe chimps is not how humans share food, which is not to say there is no theft or begging. It simply means that humans willingly share food with non-kin, while most chimps will only do so when harassed into it. Human behavior is probably at its most instinctual in young children. When compared in equivalent experimental conditions, young children happily share food with each other, while chimps avoid doing so. How did this difference evolve? It probably has its origin in the different ways in which chimps and humans find food.

Although chimps are social animals, they forage for food individually and eat on their own. This is because the fruits that form the greater part of their diet are distributed through the tree canopy and each item is too small to share. It is only on those uncommon occasions when a chimp has a very large fruit or a whole monkey carcass that it attracts the attention of others who try to beg or steal. This may also be how our tree-living ancestors foraged long ago, but when we became inhabitants of the plains of Africa, we were after much bigger game. Our eyes have been bigger than our stomachs ever since.

What other predator hunts prey so much larger than itself that it can live inside the remains of the carcass, as *Homo sapiens* once did on the mammoth steppe? Hunting big animals is only possible when people cooperate with one another. If the menagerie of giant animals painted by Paleolithic cave artists leaves any doubt that humans were social hunters, the handprints stenciled on the same cave walls must dispel this with a convincing show of hands.

Hunting large prey had stupendous consequences for the evolution of human sociality. Not only do large prey require hunters to cooperate, but they also reward the hunters for doing so with a prize that is big enough to feed everyone. When there is more than enough food to go around, there is no need to be possessive over the prey. Even chimps will help each other when the cost of doing so is very small. This leads to a hypothesis about how the human propensity to cooperate in tasks and to share the rewards evolved. It is how natural selection cut the cloth of our psyche in a world where interdependence was a necessity if you wanted to eat.

When we divide a pizza among friends or spin the lazy Susan in the center of the table at a Chinese restaurant, we are sharing food that was grown on a farm, but our meals have a much deeper evolutionary origin. Our very habit of eating together and the cooperation required to run that farm and that restaurant are all built upon an ancient legacy of communal hunting. Farms and restaurants are often family enterprises, a reminder that there is also kin selection in the weave of our psychological fabric. Reciprocation has also visibly marked the cloth.

If you have ever tried unsuccessfully to catch the eye of a waiter in a restaurant and eventually given up, as I have, then it may come as small comfort to you to know that a chimp could not do any better. Humans are extraordinarily sensitive to the regard of others. If you gaze at someone, they will notice this even if you are at the edge of their field of vision. This is possible because the whites of your eyes contrasted with the black pupils at their centers make the direction in which you are looking clearly visible to others. Chimp eyes have no whites and so they cannot tell so easily, nor quite possibly do they care, who is looking at them. Perhaps as you suspected, that waiter has not failed to notice you—he just doesn't want to see you. Or else he is a chimp.

Human eyes are not designed by evolution just to look, but also to be seen looking. We use them to signal that we are watching others. Why is this of evolutionary advantage? A hypothesis that is supported by experimental evidence is that when a social bargain exists, watching the other party keeps them honest. This is so strong and unconscious an effect that even a photograph of a pair of eyes can change behavior. Such a photograph placed experimentally over an honesty box in a university coffee room trebled the amount of money collected from coffee drinkers, compared to the amount deposited into the same box when a socially neutral photograph of flowers was displayed instead. Do try this experiment at home.

The coffee money experiment—and others like it that demonstrate a similar effect of being watched on other prosocial behaviors like not littering or drivers yielding to pedestrians at road crossings—show that there is a clear advantage to the viewer of being seen to be watching. But why do those who are being watched respond in this way? What advantage can there be in being seen to obey the social rules in public, even if you are prepared to ignore them in private? The answer seems to be that what others think of you is important. Or, as deceitful Iago says to Othello in Shakespeare's play:

> Good name in man and woman, dear my lord,
> Is the immediate jewel of their souls.
> Who steals my purse steals trash; 'tis something, nothing;
> 'Twas mine, 'tis his, and has been slave to thousands;
> But he that filches from me my good name
> Robs me of that which not enriches him,
> And makes me poor indeed.

Reputation is all. It is the currency used to underwrite all the social relationships that depend upon trust for their success. The reason, as Iago says, that reputation is more valuable than money is that it affects all relationships, including the most important ones like that between Othello and his wife, Desdemona, which is at the center of the play. A biologist watching the play might say that by sullying Desdemona's reputation and sowing the seeds of doubt about her faithfulness in the mind of her husband, Iago renders the fitness of the couple to zero. Iago murders Desdemona, though she dies at the hand of Othello. Othello is then tortured by remorse, much to Iago's satisfaction. Reputation, like Iago, operates indirectly upon relationships, but with no less force than if it were direct.

Reputation is a social asset that has many of the properties of an economic one. It must be earned, it can be lost, and it can be traded. One interpretation of the value of a feast to the person who picks up the tab is that they are trading food, which they can afford far in excess of anything they could eat themselves, for increased reputation among their guests. The appetite for food can be satiated, but for many people the appetite for status cannot.

Once basic nutritional needs are satisfied, sharing food is no longer about directly maintaining fitness, but about winning social rewards that might have an indirect effect on fitness. As W. S. Gilbert, the comic librettist of the Gilbert and Sullivan operas, said of dining: "It isn't so much what's on the table that matters, as what's on the chairs." Of course, if you want to impress the occupants of the chairs, what is on the table may actually matter quite a bit. Kings, emperors, and plutocrats have vied with each other down the centuries in the extravagance of their feasts.

In 63 BCE Servilius Rullus, one of the richest men in Rome, organized a banquet in honor of Cicero, who was then consul of the city-state. The first course of delicacies was so wonderful that the diners broke into spontaneous applause. Then the cook appeared leading four Ethiopian slaves bearing a huge silver dish on which was arranged a gigantic wild boar with baskets of dates suspended from his tusks, surrounded by baby boars made of pastry. The dish was set down while the guests, who were no doubt salivating in anticipation, looked on in hushed silence. The boar was carved open, revealing a second entire boar inside and inside this was a third. With each stroke of the carving knife, a smaller and smaller animal was uncovered until a tiny bird formed the diminutive finale.

The Boar *á la Troyenne*—as French gastronomes later named it in reference to the Trojan horse of Greek mythology—became such a sensation in Rome that households that once hesitated to serve a large boar of any description now commonly went the whole Trojan hog. No sooner had this become common, than Roman hosts began to up the ante, serving three, four, then eight, and finally twenty Boar *á la Troyenne* at a single dinner.

Two thousand years later, the fad for engastration—as chefs call it now—created the turducken: a turkey stuffed with a duck stuffed with a chicken. (Whoever coined this unfortunate name for the dish failed to recognize its scatological first syllable.) Then began the inevitable engastration escalation. English chef Hugh Fearnley-Whittingstall created a ten-bird roast on his TV show in 2005. An eighteen-pound turkey was stuffed with a goose, a duck, a mallard, a guinea fowl, a chicken, a pheasant, a partridge, a pigeon, and a woodcock. Two years later, a farm store in Devon, England, began selling a twelve-bird roast, symbolizing the twelve days of Christmas. This was big enough to serve 125 people.

The fashion for engastration illustrates how appetite for food gives way to an appetite for status once food is in surplus. The three-bird roast will certainly satisfy hunger, but evidently neither three birds nor three hogs can satisfy the desire for status because that desire is fundamentally insatiable. Hunger is satiated by negative feedback in the regulatory circuit that controls it. The hormones that stimulate hunger are switched off by eating. In contrast, the human regard for status—perhaps originating in attention to how the proceeds of the Paleolithic hunt were shared—creates another kind of circuit. This is a network of social interactions that is prone to positive feedback.

It is positive feedback that causes an audio amplifier to scream when the gain is turned up too high. Positive feedback in social networks can similarly cause them to go haywire. My three-bird roast raises my status among my dinner guests, who then feel the need to reciprocate. When everybody is serving three-bird roasts, I have become like everyone else, so I go one better and show off with a four-bird roast. Four-bird roasts become the new norm, and so I have to go one better. But no, dammit, I'm going ten better!

Positive feedback always tends to overshoot the bounds of normality into excess. An example of status-seeking getting out of control occurred among Native Americans in a tribe of the Pacific Northwest who traditionally held potlatch feasts of competitive gift-giving. The "potluck" dinner, where all guests contribute something to the meal, is derived from the term *potlatch*, but the original feast was a far cry from its modern namesake. The idea of a potlatch feast was to earn status in the eyes of your rivals by an ostentatious display of wealth and generosity. Aristocrats of the tribe earned extra titles and rank by inviting rivals to a potlatch at which the guests were showered with gifts. These included blankets, fish, sea-otter pelts, canoes, and sheets of beaten copper decorated with figures that were made expressly to be given away.

Guests who at some future potlatch did not return gifts of even greater value were humiliated. This ritual eventually escalated into the wanton destruction of wealth when chiefs would throw valuables such as blankets and canoes onto a rival's fire in order to earn kudos and place him in reputational debt. In some feast houses, the fire was kept burning fiercely by a continuous stream of precious oil pouring from a carved figure in the ceiling. Guests pretended not to notice the scorching heat and lost skin in order not to lose face. The ultimate sign of a successful potlatch was when the host's feast house burned to the ground.

Competitive gift-giving on this scale may seem irrational, but the potlatch is not a unique phenomenon and it only occurs where food is in excess. A similar practice emerged in New Guinea after the sweet potato arrived there, creating a food surplus. Potlatching ceased in the Pacific Northwest in times of food scarcity. Food and reputation are both powerful and interdependent. Even in societies where no one goes hungry and social reputation is traded freely with power, wealth, and sex, it is worth asking how reputation got to be so important. The answer, at least in part, is through cooperating in the hunt and sharing in the feast.

If it is correct, the hypothesis that we evolved cooperative behavior through mutual dependence in the hunting of large animals explains a great deal more than table manners. All group activities from sports to worship and war, every lofty political ideal based upon community, nation, or equality, as well as the institutions of democracy and the rule of law that support these ultimately derive from the ancient desire for a fair stake in a good steak.

# 14

## *Future Food*

"What shall we eat tomorrow?" is the question that everyone responsible for putting food on the table asks themselves every day, but looking much further into the future, what do we see? The future evolution of food will be dominated by two challenges: the growth of the human population and global climate change. The first of these challenges is not a new one, but climate change is going to make feeding the projected human population of 10 billion a whole lot harder. Increased temperatures, altered patterns of rainfall, more frequent drought, and eventually sea-level rise will all threaten food security unless we adapt food production systems and our crops themselves. Furthermore, agriculture as currently practiced is exacerbating climate change because it is responsible for a significant share of greenhouse gas emissions. We must therefore not only feed more people in future, but do so sustainably as well.

These are big issues with their genesis in the history of how we and our food evolved. It was the invention of agriculture in the Neolithic that gave the human population its huge kick upward. In the last 250 years, population growth has been supported by the global spread of staple crops such as wheat, potato, corn, and cassava. It could therefore be said that evolution, in the form of plant and animal breeding, is at least partly responsible for the challenges we face, but it is also essential to the solution. The

Australian poet A. D. Hope (1907–2000) summarizes some of the history in verse. It begins with hunting:

> No hunter of the Age of Fable
> Had need to buckle in his belt;
> More game than he was ever able
> To take ran wild upon the veldt;
> Each night with roast he stocked his table,
> Then procreated on the pelt.
> And that is how, of course, there came
> At last to be more men than game.

But don't worry, says the poet, for then farming was invented:

> No matter: man's invention can
> Snatch triumph from his worst mistakes.
> Soon cuts of beef and pork began
> To take the place of feral steaks

But it's going to end badly because

> Effects of over-population
> Converge, no matter where you start;
> The economics of inflation
> Follows the same curve on the chart

This is the argument that the Reverend Thomas Malthus (1766–1834) advanced in his famous book *An Essay on the Principle of Population.* He said that while population has the capacity to increase geometrically (e.g., 1, 2, 4, 8, 16 . . .), the best that we can expect of technology is that it will improve the food supply arithmetically (e.g., 1, 2, 3, 4, 5 . . .). The result of this disparity will be that population will continually press upon the food supply and only misery can result. Or, in the verse of A. D. Hope:

> And breeding as he does unchecked
> By Nature, Law or Common Caution,

No cornucopia can expect
To pour forth plenty in proportion,
Nor human skills for long perfect
New means to eke his dwindling portion

Overpopulation was of major public concern in the 1960s and '70s when Hope was writing. Two key books at the time were Paul Ehrlich's *The Population Bomb* (1968) and the Club of Rome's report *Limits to Growth* (1972), both of which predicted imminent disaster. The cause for concern was real enough, though the predictions were not borne out. Between 1960 and 1980, the global population increased by 50 percent from 3 billion people to 4.5 billion, but the food supply kept pace with this. To paraphrase Hope, the cornucopia poured forth in proportion to the geometrically increasing population. The reason it was able to do so, against expectation, was that the green revolution in agriculture multiplied the yields of the major cereal crops—wheat, rice, and corn—by 50 percent or more. Evolution itself is the cornucopia, firmly under the direction of plant and animal breeders. The urgent question now is, when there are 10 billion of us, can the cornucopia cope?

Before the green revolution, cereal plants typically had tall, spindly stems. These were liable to fall over before harvest, especially when fertilizer was applied to increase yield. These plants put more energy into leaves and stems than into their seeds, also limiting their yield. Tall and leafy stems were an evolutionary legacy from the wild where natural selection favors plants that are tall enough to avoid being over-topped and shaded by their neighbors. Farmers found uses for the long straw made by old wheat varieties. The straw hats that were the summer fashion in male headgear from the 1860s into the 1920s were made from this by-product.

In all three major cereals, the success of the green revolution was due to reducing the length of the plant stem, making it thicker and better able to support plants that had bigger seed heads. At a plant breeding laboratory in Mexico, Norman Borlaug crossed traditional wheat varieties with a dwarf variety from Japan, producing new semi-dwarf wheat varieties that were sturdy, disease resistant, and responded efficiently to nitrogen fertilizer. The introduction of these varieties across the developing world dramatically increased yields, making India, where Ehrlich had predicted imminent famine, self-sufficient in wheat. Similar breeding programs produced green revolutions in rice and corn, with equally dramatic effects upon the food supply. The green revolution not only improved food security, but by increasing the yield obtained from existing farmland, it protected an estimated 18–27 million hectares of natural habitat from conversion to agriculture.

Borlaug, the father of the green revolution, was awarded the Nobel Peace Prize in 1970, but in his Nobel Lecture he cautioned: "The green revolution has won a temporary success in man's war against hunger and deprivation; it has given man a breathing space. If fully implemented, the revolution can provide sufficient food for sustenance during the next three decades. But the frightening power of human reproduction must also be curbed; otherwise the success of the green revolution will be ephemeral only."

The gains of the green revolution have now been fully realized in many places, and crop yields have begun to level off. This leaves an anticipated gap between what the varieties grown now are producing and the yields per hectare that will be required to meet the needs of a population of 10 billion people by mid-century. By one estimate, closing the gap between what is grown now and what we shall need in order to be able to feed everyone in 2050 will require at least another 50 percent increase in crop yields. This is equivalent to raising average yields, globally, to the level of the record yields that can be achieved today. A 50 percent average yield improvement is probably achievable on current trends, but there are other, much higher estimates of future food demand that require crop yields to double. Doubling cereal yields cannot be achieved on current trends or business as usual.

Dramatically increasing cereal yields is, of course, not the only way to balance the future supply and demand for food. Concentrating only on the supply side of the equation could be called a technical fix for a problem that is social as well as scientific. Social solutions address the level of food demand and involve such things as lowering the rate of human population increase through birth control, reducing food waste, and lowering the demand for cereals that are used for animal feed by persuading people in developed countries to eat less meat. While these are all desirable developments in their own right, it would be gambling with our future to depend upon them alone, and so plant scientists argue that we need a second green revolution.

The scientific challenges of a second green revolution are different from those of the first. The challenge of the first revolution could be characterized as breeding new plant varieties that were better adapted to industrialized agriculture. It successfully produced disease-resistant crops that yielded more when fertilized and irrigated. Plant breeders working on the next green revolution are up against a more complicated set of barriers to higher yields. These include, for example, improving the salt tolerance of crops so that they can grow in soils that have become saline through past irrigation practice, improving crop resistance to drought and higher temperatures, and combating ever-evolving pests and diseases.

While the challenges for a second green revolution are greater than those of the first, the genetic tools at our disposal are enormously advanced over those available to Borlaug and the crop breeders of the 1950s and '60s. Gene sequences now exist for at least fifty crops, making it possible, for example, to identify the precise genetic mutations responsible for the dwarfing traits that were so crucial to the green revolution. A gene has been discovered in one of the ancestors of wheat that improves the salt tolerance of the plant, offering the potential for a rapid improvement in its adaptation to saline soils.

Most ambitiously of all, the fundamental mechanism of photosynthesis—the process by which plants use the energy of sunlight to capture $CO_2$ and turn it into glucose—is now sufficiently well understood that it is feasible to improve it significantly through genetic engineering. This could raise the yield of crop plants, provided that enough fertilizer and water could be supplied. There are, of course, those who oppose genetic engineering; for example, in 2015 in Scotland, where I live, the government banned the raising or cultivating of genetically modified organisms (GMOs) in order to be able to brand this corner of northern Europe "GM Free."

Growing GM crops is heavily regulated in the European Union and at the time of writing little practiced. When Romania joined the EU in 2007, farmers had to cease growing GM varieties of soybean, yields dropped as a consequence, and the crop became unprofitable. Romania had been an exporter of soybean but now had to rely on expensive imports instead. An exception to the EU ban is GM corn, which is widely grown in Spain, where farmers find they need to spray the crop with insecticide only a tenth as often as required by conventional corn grown elsewhere in Europe.

GM corn, soya, and canola are widely grown in the United States, but there is widespread distrust of GMOs among the public, with a survey conducted in 2014 finding that 57 percent of U.S. adults believe GM food to be generally unsafe. This suggests that most consumers are at best ill-informed and at worst have been misled to fear a technology that can and does benefit people. While twenty years ago it could be claimed with some justification that genetic engineering was a new and untested technology, this is no longer the case. There have now been thousands of trials of the safety of GM crops, and there is copious evidence that GM technology does not render food unsafe to grow or to eat. Perhaps because of the weight of this evidence of safety, Greenpeace, which used to argue against GMOs on grounds of safety, now argue instead that GM crops provide either no benefits or benefit the wrong people. But on the contrary, among the real benefits of GM technology, there are now examples that have improved crop yields, reduced pesticide use, and even saved whole industries from destruction by disease.

Papaya is a fruit that is important to poor subsistence farmers throughout the Tropics. It is attacked by a virus called papaya ringspot (PRSV) that drastically reduces yield and kills trees. The virus is carried from plant to plant by aphids. There is no treatment for infected plants, so farmers growing papaya can only control it by spraying insecticide to kill the aphids that carry the virus. This is expensive, polluting, and not very effective. Attempts to breed PRSV-resistant varieties of papaya by conventional methods totally failed, and the future of this crop looked very bleak as the virus spread across the Tropics from one papaya-growing region to another. For a time the main papaya-growing area on the island of Hawai'i escaped the scourge of PRSV, but in 1992 it arrived there too. Fortunately, a totally new approach to controlling PRSV was being tested, based on GM technology. This involved taking a piece of the gene that codes for the protein coat of the virus and inserting it into the genome of papaya. Papaya modified in this way is effectively inoculated against the virus and is totally immune to it.

The 1990s were the early days for the application of GM technology, and all releases were subject to rigorous controls. One major concern raised by anti-GM activists in Hawai'i was that the viral DNA might make the papaya allergenic and dangerous to eat. Testing proved this not to be so, but in any case plant scientists argued that people consuming papaya infected with PRSV were consuming huge quantities of viral DNA without ill effect. If you are worried about consuming the DNA or viral protein from PRSV, then GM papaya should actually be your fruit of choice because it is virus-free. In fact, you need not worry either way, since the virus is destroyed in your stomach.

After a struggle for regulatory approval that nearly blocked its introduction, GM papaya was approved and has been successfully grown in Hawai'i since 1998. Despite its record of safety in Hawai'i and the fact that the GM variety saved the papaya industry there, opposition to GM has blocked its introduction in parts of the developing world where it could do the most good. Greenpeace claim that GM technology is ineffective, but the organization is itself responsible for preventing its use where it would be beneficial. In Thailand in 2004, Greenpeace activists wearing goggles and respirators destroyed field trials of GM papaya, ripping fruit from the trees and throwing them into bins marked "Biohazard."

While GM food has not harmed anyone, the irrational opposition to it has almost certainly done so. Plans for the free distribution of golden rice—a GM variety engineered to provide vitamin A to communities suffering from a deficiency that causes blindness and death—have been strenuously opposed by anti-GM activists. Poor farmers in developing countries have been denied access to GM varieties that are resistant to pests and disease. In India, activists have prevented the introduction of a GM eggplant variety engineered with the *Bt* insect-resistance gene that would protect this important vegetable crop from some of its major pests. The *Bt* gene comes from a bacterium that infects and kills caterpillars. In contrast, India does permit the cultivation of *Bt* cotton. Since its introduction, this has had both environmental and economic benefits for small farmers, who get better yields without the heavy use of pesticides. Why should the small Indian farmers who grow eggplant be denied similar benefits?

Genetically modified crops have enormous potential for use in sustainable agriculture. Pest-resistant crops such as *Bt* eggplant can reduce disease and improve yields while at the same time reducing agricultural inputs such as pesticides. Greater efficiency of water use, a real possibility through genetic engineering, would reduce one of the biggest environmental impacts of agriculture. It is tragic that GM has been demonized so that consumers are misled to believe that a technology that can do good must do harm.

Aside from the damage that ignoring scientific evidence does to livelihoods and the environment, well-meaning campaigns directed against the wrong target undermine the credibility of those who espouse important environmental causes. How can you trust people and organizations that do not respect scientific evidence? This realization caused one high-profile campaigner, Mark Lynas, to change his mind on GM crops. In 2015 he wrote in the *New York Times*:

> A lifelong environmentalist, I opposed genetically modified foods in the past. Fifteen years ago, I even participated in vandalizing field trials in Britain. Then I changed my mind. After writing two books on the science of climate change, I decided I could no longer continue taking a pro-science position on global warming and an anti-science position on GMOs. There is an equivalent level of scientific consensus on both issues, I realized, that climate change is real and genetically modified foods are safe. I could not defend the expert consensus on one issue while opposing it on the other.

The question of GM crops is an evolutionary issue in four ways. First, despite current opposition, genetic modification will determine the future evolution of our crops. This is how our food will evolve. It remains to be seen whether other opponents will have the moral courage of Lynas to publicly admit that they were mistaken. However, the steam will be taken out of the controversy as the realization spreads that GMOs are impossible to define in a manner that clearly distinguishes them from the crops and animals that we have genetically modified over millennia of domestication. The reason for this is the second way in which GM is an evolutionary issue: Nature herself is the original genetic engineer.

One of the key discoveries of the genomic revolution is that genes naturally cross the species barrier through horizontal gene transfer (HGT; chapter 11). Viruses and some bacteria are the main agents of HGT, both in the lab and in nature. A soil bacterium called *Rhizobium radiobacter* that infects the roots of a wide range of broad-leaved plants transfers some of its DNA into plant cells during the infection process. Since its discovery in the late 1970s, this natural process in *R. radiobacter* has been widely used as a delivery mechanism to transfer genes such as the *Bt* gene into crop plants.

The third reason that GM is an evolutionary issue is that natural selection has already evolved and tested most of the technologies, so in using them we are working with nature and not against her. For example, the genome of the domesticated sweet potato contains genes that originate from a bacterium like *Rhizobium radiobacter*. These appear to have been acquired during the domestication process since the DNA sequences in question are not present in the wild relatives of the crop. What function or functions these genes perform in sweet potato is not as yet known, though they are likely to confer traits beneficial to the use or storage of the crop.

The latest GM technology acquired from nature and certainly the most revolutionary to date is the system known as CRISPR-Cas9. This is a genome editing system found in bacteria that gives them adaptive immunity to viruses. It enables a bacterial cell to recognize viral DNA sequences that have inserted themselves into the bacterial chromosome, to snip them out, and then to repair the break. In the lab, any short DNA sequence can be targeted for editing by programming Cas9 with its corresponding RNA template.

In effect, the CRISPR-Cas9 system can be used to edit DNA sequences in the same way that you would use the search-and-replace function in a word-processing program to edit a document. By introducing into an animal or plant cell the bacterial genes that make the components of the CRISPR-Cas9 system, the DNA sequence of the recipient cell becomes editable. The potential impact of this new tool for genome editing on medicine and agriculture is difficult to overstate. To give just two examples, in medicine it will enable faulty genes that cause inherited disease such as cystic fibrosis to be repaired. In plants, CRISPR-Cas9 has already been used to alter a gene that makes bread wheat susceptible to mildew, rendering it resistant to a devastating disease that threatens food security. The triple genome of bread wheat makes breeding for mildew resistance difficult to achieve by means other than genome editing.

GM technologies are not unnatural or untested, but we should not be complacent about the awesome power that they give us or, conversely, expect them to solve every problem. This brings us to the fourth way in which GM is an evolutionary question: pests can evolve resistance to GM technologies designed to defeat them. So, for example, weeds have evolved resistance to glyphosate, the herbicide that is used on GM crops that have been engineered to tolerate this treatment. Insects have also evolved resistance to the *Bt* toxin that is produced by GM crops engineered with the *Bt* gene.

These examples only tell us what we already know, which is that evolution is ongoing and ever present; they do not mean, as some anti-GM activists have claimed, that GM technology is a failure. The evolution of resistance in pests can be limited through integrated pest management. This uses a variety of means, of which GM is potentially only one, to produce the best yields in a sustainable way. For example, GM varieties can be included in a crop rotation. This is the traditional practice of cycling a field through a variety of crops over a period of several years to maintain soil fertility and to control the buildup of pests.

All forms of plant and animal breeding, GM included, can potentially have unintended consequences. This is not because GM is inherently more risky than other breeding technologies, but because novelty always involves risk. However, the novelties that are associated with the greatest threats to health and the environment are not GM crops or domesticated species, but wild species such as the Argentine ant, the zebra mussel, or the kudzu vine that have each done untold damage when introduced outside their natural ranges. Nothing we do, or fail to do, is without risk, so all risks must be judged in proportion. At present the risks of GMOs tend to be vastly overrated, while the potential benefits for sustainable food production are far too little appreciated.

And so we conclude our dinner with Darwin, and this book can now claim its place in the library alongside *The Complete Idiot's Guide to Smoking Foods*, *Bubbles in Food*, and *A Diet of Tripe* (chapter 1). If you have dined with me through all the courses, you may have noticed that evolution and cooking are alike in a fundamental way. Innovations in evolutionary history, even huge ones like the origin of mammals or birds, are assembled from pre-existing characteristics. Lactation was present in the predecessors of mammals, just as eggs, feathers, and some form of flight were present in the ancestors of birds. Before cereal agriculture developed in the Fertile Crescent, people were gathering the seeds of wild grasses for 20,000 years. In genetic terms, selection—whether natural or artificial—works on existing variation.

How is this like cooking? Well, first it's how cooking evolved too, but it's also how cooks work. You use what evolution has provided and what you have in the cupboard or the market. Is there a lesson here? Yes, I think so. Evolution is all about the potential of its ingredients, and so is good cooking. These are facts ignored by anyone who wants to tell you that our evolution dictates that we should restrict our diet according to some fanciful notion of what we ate in the Paleolithic. Our evolutionary history has indeed shaped our dietary capabilities, but it has broadened rather than narrowed them. We have survived the waxing and waning of ice sheets and deserts, and then thrived, multiplied, and occupied every continent because we are adaptable, intelligent omnivores. If we were not, we would be as endangered as the giant panda that eats little other than bamboo shoots or the koala that lives on eucalyptus. Ironically, if our numbers were thus curtailed, those two species would now be less severely threatened, no doubt.

Dietary studies confirm what a comparison of the various diets of different cultures would suggest, which is that there are many ways to achieve a healthy, balanced diet and only the extremes that overdose on meat or avoid animal protein altogether tend to be problematic. For diets between these extremes, the biggest dietary threat to health is the very modern one of consuming too many calories.

With such diversity of food available for our enjoyment, one might wonder why so many authors want to convince us that evolution mandates a restricted diet. The likely answer was unwittingly provided to me by a literary agent to whom I sent the outline for this book. He told me that I should follow the pack and write an evolutionary prescription for the human diet because this is what would sell. I'd rather sell you the Brooklyn Bridge.

And finally, you might be curious as to what an actual dinner with Darwin might have been like. If instead of just inspiring this dinner, he could have shared it with us, Darwin would no doubt have been staggered to learn of the progress that we have made in genetics and what this has revealed about evolution. But sadly, Charles Darwin was afflicted by a nearly lifelong stomach complaint that meant he rarely gave or attended dinner parties. Darwin recorded in his autobiography that when he and his wife, Emma, first moved to the village of Downe in Kent to escape the clamor of early Victorian London, they did give a few dinner parties. Emma wrote to her sister about one such dinner on April 1, 1839, that was attended by John Stevens Henslow and Charles Lyell, saying, "Notwithstanding those two dead weights, viz., the greatest botanist [Henslow] and the greatest geologist in Europe [Lyell], we did very well and had no pauses," thanks to the good conversation provided by the great men's wives.

But the age of dinners with the Darwins was short-lived. As Emma wrote, Darwin's ill health soon compelled them to "give up all dinner parties and this has been somewhat of a deprivation to me, as such parties always put me in high spirits." As it happens, Emma Darwin kept a recipe book, so we do have a good idea of some of the fancier dishes served from her kitchen, but whether due to Charles's invalid stomach or the limitations of Victorian cuisine, there is little in it to inspire the modern cook. Darwin's real contribution was his discovery of the recipe for evolution.

# Acknowledgments

As ever, I am indebted to my wife, Rissa de la Paz, for her uncompromising scrutiny of the manuscript and clear-sighted vision of what makes a good book. I hope that I am getting nearer this target with each shot at the goal. I am grateful to my longtime colleague and friend Professor Caroline Pond, who knows more biology than almost anyone and who read the whole manuscript in search of blunders. Any that escaped are purely my responsibility. I thank Professor Pam Ronald at UC Davis, who provided comments on the chapter "Future Food" and gave me advance sight of chapters in the forthcoming new edition of her book, written with Raoul Adamchak, *Tomorrow's Table*. I am also grateful to Professor Sharon Strauss, who was my host in a too-brief stay at UC Davis, where I began writing this book. Finally, it is a pleasure to acknowledge some new friends, the bunch of nonfiction writers who meet once a month at the Wash Bar in Edinburgh under the banner of "Stranger than Fiction." A dozen of them read selected chapters of this book and made insightful comments, which I greatly appreciate.

## MAP SOURCES

*Map 1*. The locations of the finds and events shown in the map come from the original sources cited in chapters 2 and 3.

*Map 2.* The routes shown are based upon those given in figure 1 in S. Oppenheimer, "Out-of-Africa, the Peopling of Continents and Islands: Tracing Uniparental Gene Trees across the Map," *Philosophical Transactions of the Royal Society of London B: Biological Sciences* 367, no. 1590 (2012): 770–84. Dates were taken from Oppenheimer and more recent sources cited in chapter 3.

*Map 3.* Based upon figure 1 in D. Q. Fuller et al., "Cultivation and Domestication Had Multiple Origins: Arguments against the Core Area Hypothesis for the Origins of Agriculture in the Near East," *World Archaeology* 43, no. 4 (2011): 628–52. Additional information comes from sources cited in chapter 4.

*Map 4.* Compiled from maps of Vavilov's travels given in N. I. Vavilov, *Five Continents by Nicolai Ivanovich Vavilov*, translated from the Russian by Doris Löve (IPGRI; VIR, 1997).

*Map 5.* Based on A. A. Storey et al., "Investigating the Global Dispersal of Chickens in Prehistory Using Ancient Mitochondrial DNA Signatures," *PLOS ONE* 7, no. 7 (2012); H. Xiang, et al., "Early Holocene Chicken Domestication in Northern China," *Proceedings of the National Academy of Sciences of the United States of America* 111, no. 49 (2014): 17564–69; Y. W. Miao et al., "Chicken Domestication: An Updated Perspective Based on Mitochondrial Genomes," *Heredity (Edinburgh)* 110, no. 3 (2013): 277–82, doi:10.1038/hdy.2012.83.

*Map 6.* Based on figure 1 in M. A. Zeder, "Domestication and Early Agriculture in the Mediterranean Basin: Origins, Diffusion, and Impact," *Proceedings of the National Academy of Sciences of the United States of America* 105, no. 33 (2008): 11597–604.

# *Notes*

CHAPTER ONE

*The Complete Idiot's Guide*: T. Reader, *The Complete Idiot's Guide to Smoking Foods* (Alpha/Penguin Group, 2012).

*Bubbles in Food*: G. M. Campbell, *Bubbles in Food* (Eagan Press, 1999); G. M. Campbell et al., *Bubbles in Food 2: Novelty, Health, and Luxury* (AACC International, 2008).

*A Diet of Tripe*: T. McLaughlin, *A Diet of Tripe: The Chequered History of Food Reform* (David & Charles, 1978).

*No More Bull!*: H. F. Lyman et al., *No More Bull!: The Mad Cowboy Targets America's Worst Enemy, Our Diet* (Scribner, 2005).

*Handheld Pies*: R. Wharton and S. Billingsley, *Handheld Pies: Pint-Sized Sweets and Savories* (Chronicle Books, 2012).

Oxford Symposium: H. Saberi, ed., *Cured, Fermented and Smoked Foods: Proceedings of the Oxford Symposium on Food and Cookery, 2010* (Prospect Books, 2011).

*Twin-Screw Extrusion*: I. Hayakawa, *Food Processing by Ultra High Pressure Twin-Screw Extrusion* (Technomic Publishing, 1992).

dinosaurs also nested: D. J. Varricchio et al., "Avian Paternal Care Had Dinosaur Origin," *Science* 322, no. 5909 (2008): 1826–28, doi:10.1126/science.1163245.

France is still a global hotspot for these fossils: R. Allain and X. P. Suberbiola, "Dinosaurs of France," *Comptes Rendus Palevol* 2, no. 1 (2003): 27–44, doi:10.1016/s1631-0683(03)00002-2.

9½ tons of milk a year: USDA, *Milk Cows and Production Final Estimates, 2003–2007* (2009).

enough energy to sustain 400 people daily: O. T. Oftedal, "The Evolution of Milk Secretion and Its Ancient Origins," *Animal* 6, no. 3 (2012): 355–68, doi:10.1017/s1751731111001935.

became pseudogenes: D. Brawand et al., "Loss of Egg Yolk Genes in Mammals and the Origin of Lactation and Placentation," *PLOS Biology* (2008), doi:10.1371/journal.pbio.0060063.g001.

*An Orchard Invisible*: J. Silvertown, *An Orchard Invisible: A Natural History of Seeds* (University of Chicago Press, 2009).

## CHAPTER TWO

Cooking Animal: J. Boswell, *The Journal of a Tour to the Hebrides with Samuel Johnson, LLD* (1785), http://www.gutenberg.org/ebooks/6018 (accessed February 22, 2015).

intelligent enough: F. Warneken and A. G. Rosati, "Cognitive Capacities for Cooking in Chimpanzees," *Proceedings of the Royal Society B: Biological Sciences* 282, no. 1809 (2015), doi:10.1098/rspb.2015.0229.

erased by extinction: W. H. Kimbel and B. Villmoare, "From *Australopithecus* to *Homo*: The Transition That Wasn't," *Philosophical Transactions of the Royal Society of London, Series B: Biological Sciences* 371, no. 1698 (2016), doi:10.1098/rstb.2015.0248.

Charles Darwin deduced this: C. Darwin, *The Descent of Man, and Selection in Relation to Sex* (J. Murray, 1901).

mirror for a cover: Ibid., 242.

forensic deduction: J. Kappelman et al., "Perimortem Fractures in Lucy Suggest Mortality from Fall Out of Tall Tree," *Nature* (2016), doi:10.1038/nature19332.

wider range of environments: K. M. Stewart, "Environmental Change and Hominin Exploitation of C4-Based Resources in Wetland/Savanna Mosaics," *Journal of Human Evolution* 77 (2014): 1–16, doi:10.1016/j.jhevol.2014.10.003.

a great deal of chewing: D. Lieberman, *The Evolution of the Human Head* (Belknap Press of Harvard University Press, 2011), 434.

I asked Koko: R. Wrangham, *Catching Fire: How Cooking Made Us Human* (Profile Books, 2009). 91.

cannot have been completely vegetarian: S. P. McPherron et al., "Evidence for Stone-Tool-Assisted Consumption of Animal Tissues Before 3.39 Million Years Ago at Dikika, Ethiopia," *Nature* 466, no. 7308 (2010): 857–60, doi:10.1038/nature09248.

stone tools were manufactured: S. Harmand et al., "3.3-Million-Year-Old Stone Tools from Lomekwi 3, West Turkana, Kenya," *Nature* 521, no. 7552 (2015): 310–15, doi:10.1038/nature14464.

hominins living in Ethiopia: M. Dominguez-Rodrigo et al., "Cutmarked Bones from Pliocene Archaeological Sites, at Gona, Afar, Ethiopia: Implications for the Function of the World's Oldest Stone Tools," *Journal of Human Evolution* 48, no. 2 (2005): 109–21, doi:10.1016/j.jhevol.2004.09.004.

pushing the origin of *H. habilis*: F. Spoor et al., "Reconstructed *Homo habilis* Type OH 7 Suggests Deep-Rooted Species Diversity in Early Homo," *Nature* 519, no. 7541 (2015): 83–86, doi:10.1038/nature14224.

with the same vigor as Lucy: Lieberman, *The Evolution of the Human Head*, 503.

a new fossil jaw: B. Villmoore et al., "Early Homo at 2.8 Ma from Ledi-Geraru, Afar, Ethiopia," *Science* (2015), doi:10.1126/science.aaa1343.

proportions are similar to our own: C. Ruff, "Variation in Human Body Size and Shape," *Annual Review of Anthropology* 31 (2002): 211–32, doi:10.1146/annurev.anthro.31.040402.085407.

half as much chewing: Lieberman, *The Evolution of the Human Head*.

hippo, rhino, and crocodile: D. R. Braun et al., "Early Hominin Diet Included Diverse Terrestrial and Aquatic Animals 1.95 Ma in East Turkana, Kenya," *Proceedings of the National Academy of Sciences of the United States of America* 107, no. 22 (2010): 10002–7, doi:10.1073/pnas.1002181107.

rabbit starvation: S. Bilsborough and N. Mann, "A Review of Issues of Dietary Protein Intake in Humans," *International Journal of Sport Nutrition and Exercise Metabolism* 16, no. 2 (2006): 129–52.

hunter-gatherers: A. Strohle and A. Hahn, "Diets of Modern Hunter-Gatherers Vary Substantially in Their Carbohydrate Content Depending on Ecoenvironments: Results from an Ethnographic Analysis," *Nutrition Research* 31, no. 6 (2011): 429–35, doi:10.1016/j.nutres.2011.05.003.

tropical grasses or sedges: J. Lee-Thorp et al., "Isotopic Evidence for an Early Shift to C$_4$ Resources by Pliocene Hominins in Chad," *Proceedings of the National Academy of Sciences of the United States of America* 109, no. 50 (2012): 20369–72, doi:10.1073/pnas.1204209109.

ancient Egypt: D. Zohary et al., *Domestication of Plants in the Old World: The Origin and Spread of Domesticated Plants in South-West Asia, Europe, and the Mediterranean Basin* (Oxford University Press, 2012), 158.

1,900 plants: M. E. Tumbleson and T. Kommedahl, "Reproductive Potential of *Cyperus esculentus* by Tubers," *Weeds* 9, no. 4 (1961): 646–53, doi:10.2307/4040817.

experimental flake tools: C. Lemorini et al., "Old Stones' Song: Use-Wear Experiments and Analysis of the Oldowan Quartz and Quartzite Assemblage from Kanjera South (Kenya)," *Journal of Human Evolution* 72 (2014): 10–25, doi:10.1016/j.jhevol.2014.03.002.

most complete early human skull: D. Lordkipanidze et al., "A Complete Skull from Dmanisi, Georgia, and the Evolutionary Biology of Early Homo," *Science* 342 (2013): 326–31.

Elephants were hunted: M. Ben-Dor et al., "Man the Fat Hunter: The Demise of *Homo erectus* and the Emergence of a New Hominin Lineage in the Middle Pleistocene (ca. 400 kyr) Levant," *PLOS ONE* 6, no. 12 (2011), doi:10.1371/journal.pone.0028689.

local elephant species went extinct: T. Surovell et al., "Global Archaeological Evidence for Proboscidean Overkill," *Proceedings of the National Academy of Sciences of the United States of America* 102, no. 17 (2005): 6231–36, doi:10.1073/pnas.0501947102.

first cookouts happened: S. E. Bentsen, "Using Pyrotechnology: Fire-Related Features and Activities with a Focus on the African Middle Stone Age," *Journal of Archaeological Research* 22, no. 2 (2014): 141–75, doi:10.1007/s10814-013-9069-x.

biological as well as paleoarchaeological evidence: J. A. J. Gowlett and R. W. Wrangham, "Earliest Fire in Africa: Towards the Convergence of Archaeological Evidence and the Cooking Hypothesis," *Azania-Archaeological Research in Africa* 48, no. 1 (2013): 5–30, doi:10.1080/0067270x.2012.756754.

*How Cooking Made Us Human*: Wrangham, *Catching Fire*.

a gene called *MHY16*: G. H. Perry et al., "Insights into Hominin Phenotypic and Dietary Evolution from Ancient DNA Sequence Data," *Journal of Human Evolution* 79 (2015): 55–63, doi:10.1016/j.jhevol.2014.10.018.

Cooking increases the digestibility of food: R. N. Carmody and R. W. Wrangham, "The Energetic Significance of Cooking," *Journal of Human Evolution* 57, no. 4 (2009): 379–91, doi:10.1016/j.jhevol.2009.02.011.

Meat and fat: R. N. Carmody et al., "Energetic Consequences of Thermal and Nonthermal Food Processing," *Proceedings of the National Academy of Sciences of the United States of America* 108, no. 48 (2011): 19199–203, doi:10.1073/pnas.1112128108; E. E. Groopman et al., "Cooking Increases Net Energy Gain from a Lipid-Rich Food," *American Journal of Physical Anthropology* 156, no. 1 (2015): 11–18, doi:10.1002/ajpa.22622.

size is not everything: G. Roth and U. Dicke, "Evolution of the Brain and Intelligence," *Trends in Cognitive Sciences* 9, no. 5 (2005): 250–57, doi:10.1016/j.tics.2005.03.005.

Most of this energy: J. J. Harris et al., "Synaptic Energy Use and Supply," *Neuron* 75, no. 5 (2012): 762–77, doi:10.1016/j.neuron.2012.08.019.

economizing on guts: L. C. Aiello and P. Wheeler, "The Expensive Tissue Hypothesis: The Brain and the Digestive System in Human and Primate Evolution," *Current Anthropology* 36, no. 2 (1995): 199–221, doi:10.1086/204350; A. Navarrete et al., "Energetics and the Evolution of Human Brain Size," *Nature* 480, no. 7375 (2011): 91–93, doi:10.1038/nature10629.

27 percent higher: H. Pontzer et al., "Metabolic Acceleration and the Evolution of Human Brain Size and Life History," *Nature* 533, no. 7603 (2016): 390–92, doi:10.1038/nature17654.

our brains grew: Wrangham, *Catching Fire*.

*H. heidelbergensis*: L. T. Buck and C. B. Stringer, "*Homo heidelbergensis*," *Current Biology* 24, no. 6 (2014): R214–15, doi:10.1016/j.cub.2013.12.048.

fire whenever they needed it: Bentsen, "Using Pyrotechnology"; N. Goren-Inbar et al., "Evidence of Hominin Control of Fire at Gesher Benot Ya'aqov, Israel," *Science* 304, no. 5671 (2004): 725–27, doi:10.1126/science.1095443.

made of spruce wood: H. Thieme, "Lower Palaeolithic Hunting Spears from Germany," *Nature* 385, no. 6619 (1997): 807–10, doi:10.1038/385807a0.

hunted and butchered horses: T. van Kolfschoten, "The Palaeolithic Locality Schöningen (Germany): A Review of the Mammalian Record," *Quaternary International* 326–27 (2014): 469–80, doi:10.1016/j.quaint.2013.11.006.

their meals: M. Balter, "The Killing Ground," *Science* 344, no. 6188 (2014): 1080–83.

an extinct cousin: D. Reich et al., "Genetic History of an Archaic Hominin Group from Denisova Cave in Siberia," *Nature* 468, no. 7327 (2010): 1053–60, doi:10.1038/nature09710.

telltales of an encounter: D. Reich et al., "Denisova Admixture and the First Modern Human Dispersals into Southeast Asia and Oceania," *American Journal of Human Genetics* 89, no. 4 (2011): 516–28, doi:10.1016/j.ajhg.2011.09.005.

a redhead: C. Lalueza-Fox et al., "A Melanocortin 1 Receptor Allele Suggests Varying Pigmentation among Neanderthals," *Science* 318, no. 5855 (2007): 1453–55, doi:10.1126/science.1147417.

common ancestor: K. Prüfer et al., "The Complete Genome Sequence of a Neanderthal from the Altai Mountains," *Nature* 505, no. 7481 (2014): 43–49, doi:10.1038/nature12886.

40,000 years ago: T. Higham et al., "The Timing and Spatiotemporal Patterning of Neanderthal Disappearance," *Nature* 512, no. 7514 (2014): 306–9, doi:10.1038/nature13621.

slightly larger brains: A. W. Froehle and S. E. Churchill, "Energetic Competition between Neandertals and Anatomically Modern Humans," *PaleoAnthropology* (2009): 96–116.

Neanderthal feces: A. Sistiaga et al., "The Neanderthal Meal: A New Perspective Using Faecal Biomarkers," *PLOS ONE* 9, no. 6 (2014), doi:10.1371/journal.pone.0101045.

smoke particles: A. G. Henry et al., "Microfossils in Calculus Demonstrate Consumption of Plants and Cooked Foods in Neanderthal Diets (Shanidar III, Iraq; Spy I and II, Belgium)," *Proceedings of the National Academy of Sciences of the United States of America* 108, no. 2 (2011): 486–91, doi:10.1073/pnas.1016868108.

Mount Carmel: E. Lev et al., "Mousterian Vegetal Food in Kebara Cave, Mt. Carmel," *Journal of Archaeological Science* 32, no. 3 (2005): 475–84, doi:10.1016/j.jas.2004.11.006.

not very different: A. G. Henry et al., "Plant Foods and the Dietary Ecology of Neanderthals and Early Modern Humans," *Journal of Human Evolution* 69 (2014): 44–54, doi:10.1016/j.jhevol.2013.12.014.

shellfish: I. Gutierrez-Zugasti et al., "The Role of Shellfish in Hunter-Gatherer Societies during the Early Upper Palaeolithic: A View from El Cuco Rockshelter, Northern Spain," *Journal of Anthropological Archaeology* 32, no. 2 (2013): 242–56, doi:10.1016/j.jaa.2013.03.001; D. C. Salazar-Garcia et al., "Neanderthal Diets in Central and Southeastern Mediterranean Iberia," *Quaternary International* 318 (2013): 3–18, doi:10.1016/j.quaint.2013.06.007.

Rock doves: R. Blasco et al., "The Earliest Pigeon Fanciers," *Scientific Reports* 4, no. 5971 (2014), doi:10.1038/srep05971.

## CHAPTER THREE

Boke of Kokery: quoted in W. Sitwell, *A History of Food in 100 Recipes* (Little, Brown, 2013), 58.

many monkeys and apes: A. E. Russon et al., "Orangutan Fish Eating, Primate Aquatic Fauna Eating, and Their Implications for the Origins of Ancestral Hominin Fish Eating," *Journal of Human Evolution* 77 (2014): 50–63, doi:10.1016/j.jhevol.2014.06.007.

Mounds of discarded seashells: M. Álvarez et al., "Shell Middens as Archives of Past Environments, Human Dispersal and Specialized Resource Management," *Quaternary International* 239, nos. 1–2 (2011): 1–7, doi:10.1016/j.quaint.2010.10.025.

crucial to brain development: J. T. Brenna and S. E. Carlson, "Docosahexaenoic Acid and Human Brain Development: Evidence That a Dietary Supply Is Needed for Optimal Development," *Journal of Human Evolution* 77 (2014): 99–106, doi:10.1016/j.jhevol.2014.02.017; S. C. Cunnane and M. A. Crawford, "Energetic and Nutritional Constraints on Infant Brain Development: Implications for Brain Expansion during Human Evolution," *Journal of Human Evolution* 77 (2014): 88–98, doi:10.1016/j.jhevol.2014.05.001.

dined on species of shellfish: C. W. Marean et al., "Early Human Use of Marine Resources and Pigment in South Africa during the Middle Pleistocene," *Nature* 449, no. 7164 (2007): 905–8, doi:10.1038/nature06204.

most of Africa too inhospitable: C. W. Marean, "When the Sea Saved Humanity," *Scientific American* 303, no. 2 (2010): 54–61, doi:10.1038/scientificamerican0810-54; C. W. Marean, "Pinnacle Point Cave 13B (Western Cape Province, South Africa) in Context: The Cape Floral Kingdom, Shellfish, and Modern Human Origins," *Journal of Human Evolution* 59, nos. 3–4 (2010): 425–43, doi:10.1016/j.jhevol.2010.07.011.

the coast of Eritrea: R. C. Walter et al., "Early Human Occupation of the Red Sea Coast of Eritrea during the Last Interglacial," *Nature* 405, no. 6782 (2000): 65–69, doi:10.1038/35011048.

all the way to China: W. Liu et al., "The Earliest Unequivocally Modern Humans in Southern China," *Nature* 526, no. 7575 (2015): 696–99, doi:10.1038/nature15696.

roasted in a fire: M. Cortes-Sanchez et al., "Earliest Known Use of Marine Resources by Neanderthals," *PLOS ONE* 6, no. 9 (2011), doi:10.1371/journal.pone.0024026.

*H. sapiens* from North Africa traveled eastward: E. A. A. Garcea, "Successes and Failures of Human Dispersals from North Africa," *Quaternary International* 270 (2012): 119–28, doi:10.1016/j.quaint.2011.06.034.

pressure of a growing population: P. Mellars, "Why Did Modern Human Populations Disperse from Africa ca. 60,000 Years Ago? A New Model," *Proceedings of the National Academy of Sciences of the United States of America* 103, no. 25 (2006): 9381–86, doi:10.1073/pnas.0510792103.

recorded thus in our genes: S. Oppenheimer, "Out-of-Africa, the Peopling of Continents and Islands: Tracing Uniparental Gene Trees across the Map," *Philosophical Transactions of the Royal Society of London, Series B: Biological Sciences* 367, no. 1590 (2012): 770–84, doi:10.1098/rstb.2011.0306.

Africa is richly diverse: S. A. Tishkoff et al., "The Genetic Structure and History of Africans and African Americans," *Science* 324, no. 5930 (2009): 1035–44, doi:10.1126/science.1172257.

genetic diversity we lost: S. Ramachandran et al., "Support from the Relationship of Genetic and Geographic Distance in Human Populations for a Serial Founder Effect Originating in Africa," *Proceedings of the National Academy of Sciences of the United States of America* 102, no. 44 (2005): 15942–47, doi:10.1073/pnas.0507611102.

45,000 years ago: T. D. Weaver, "Tracing the Paths of Modern Humans from Africa," *Proceedings of the National Academy of Sciences of the United States of America* 111 (2014): 7170–71.

the coast had become ice-free: E. J. Dixon, "Late Pleistocene Colonization of North America from Northeast Asia: New Insights from Large-Scale Paleogeographic Reconstructions," *Quaternary International* 285 (2013): 57–67, doi:10.1016/j.quaint.2011.02.027.

All native Americans: T. Goebel et al., "The Late Pleistocene Dispersal of Modern Humans in the Americas," *Science* 319, no. 5869 (2008): 1497–502, doi:10.1126/science.1153569.

butchered mastodon bones: E. Marris, "Underwater Archaeologists Unearth Ancient Butchering Site," *Nature* (May 13, 2016), doi:10.1038/nature.2016.19913.

Pacific coast: J. M. Erlandson and T. J. Braje, "From Asia to the Americas by Boat? Paleogeography, Paleoecology, and Stemmed Points of the Northwest Pacific," *Quaternary International* 239, nos. 1–2 (2011): 28–37, doi:10.1016/j.quaint.2011.02.030.

reaching Chile: T. D. Dillehay, *Monte Verde, a Late Pleistocene Settlement in Chile: The Archaeological Context and Interpretation* (Smithsonian Institution Press, 1997).

Tierra del Fuego: A. Prieto et al., "The Peopling of the Fuego-Patagonian Fjords by Littoral Hunter-Gatherers after the Mid-Holocene H1 Eruption of Hudson Volcano," *Quaternary International* 317 (2013): 3–13, doi:10.1016/j.quaint.2013.06.024.

living chiefly upon shellfish: C. Darwin, *The Voyage of HMS Beagle* (Folio Society, 1860), chap. 10.

Archaeological excavations: L. A. Orquera et al., "Littoral Adaptation at the Southern End of South America," *Quaternary International* 239, nos. 1–2 (2011): 61–69, doi:10.1016/j.quaint.2011.02.032.

## CHAPTER FOUR

is an ancestor: D. Zohary et al., *Domestication of Plants in the Old World: The Origin and Spread of Domesticated Plants in South-West Asia, Europe, and the Mediterranean Basin* (Oxford University Press, 2012); P. J. Berkman et al., "Dispersion and Domestication Shaped the Genome of Bread Wheat," *Plant Biotechnology Journal* 11, no. 5 (2013): 564–71, doi:10.1111/pbi.12044.

workers as well as royalty ate wheat bread: D. Samuel, "Investigation of Ancient Egyptian Baking and Brewing Methods by Correlative Microscopy," *Science* 273, no. 5274 (1996): 488–90, doi:10.1126/science.273.5274.488; D. Samuel, "Bread Making and Social Interactions at the Amarna Workmen's Village, Egypt," *World Archaeology* 31, no. 1 (1999): 121–44.

King Nebhepetre Mentuhotep II: Model in the British Museum: http://culturalinstitute.britishmuseum.org/asset-viewer/model-from-the-tomb-of-nebhepetre-mentuhotep-ii/ygG7V06b8fjrfQ?hl=en (accessed November 19, 2016).

tooth wear of Egyptian mummies: J. E. Harris, "Dental Care," *Oxford Encyclopedia of Ancient Egypt*, vol. 1, ed. D. B. Redford (Oxford University Press, 2001): 383–85.

hieroglyphs that, deciphered, read like a comic book: http://www.osirisnet.net/tombes/nobles/antefoqer/e_antefoqer_02.htm (accessed March 12, 2014).

200 kinds of bread: J. Bottéro, *Cooking in Mesopotamia*, trans. T. L. Fagan (University of Chicago Press, 2011).

A dry climate favors the evolution of large seeds: A. T. Moles and M. Westoby, "Seedling Survival and Seed Size: A Synthesis of the Literature," *Journal of Ecology* 92, no. 3 (2004): 372–83.

wild einkorn wheat: J. R. Harlan, "Wild Wheat Harvest in Turkey," *Archaeology* 20, no. 3 (1967): 197–201.

why should anyone: J. R. Harlan and D. Zohary, "Distribution of Wild Wheats and Barley," *Science* 153, no. 3740 (1966): 1074–80, doi:10.1126/science.153.3740.1074.

domestication took thousands of years: M. D. Purugganan and D. Q. Fuller, "Archaeological Data Reveal Slow Rates of Evolution during Plant Domestication," *Evolution* 65, no. 1 (2011): 171–83, doi:10.1111/j.1558-5646.2010.01093.x.

Wild emmer and wild barley were gathered: Zohary et al., *Domestication of Plants in the Old World*.

earliest site where such remains have been discovered: Ibid.

evolving under domestication: D. Q. Fuller et al., "Moving Outside the Core Area," *Journal of Experimental Botany* 63, no. 2 (2012): 617–33, doi:10.1093/jxb/err307; P. Civan et al., "Reticulated Origin of Domesticated Emmer Wheat Supports a Dynamic Model for the Emergence of Agriculture in the Fertile Crescent," *PLOS ONE* 8, no. 11 (2013), doi:10.1371/journal.pone.0081955.

found their winters too severe: C. Darwin, *The Variation of Animals and Plants under Domestication*, vol. 1 (John Murray, 1868).

insufficient rail cars: http://www.agcanada.com/daily/statscan-shows-shockingly-large-crops-all-around (accessed March 19, 2014).

between 500,000 and 800,000 years ago: T. Marcussen et al., "Ancient Hybridizations among the Ancestral Genomes of Bread Wheat," *Science* 345, no. 6194 (2014), doi:10.1126/science.1250092.

as recently as 8,000 years ago: Zohary et al., *Domestication of Plants in the Old World*; J. Dvorak et al., "The Origin of Spelt and Free-Threshing Hexaploid Wheat," *Journal of Heredity* 103, no. 3 (2012): 426–41, doi:10.1093/jhered/esr152.

at least 230,000 years ago: Marcussen et al., "Ancient Hybridizations among the Ancestral Genomes of Bread Wheat."

evolutionary versatility of bread wheat: J. Dubcovsky and J. Dvorak, "Genome Plasticity a Key Factor in the Success of Polyploid Wheat under Domestication," *Science* 316, no. 5833 (2007): 1862–66, doi:10.1126/science.1143986.

A strain of stem rust called Ug99: R. P. Singh et al., "The Emergence of Ug99 Races of the Stem Rust Fungus Is a Threat to World Wheat Production," *Annual Review of Phytopathology* 49, no. 1 (2011): 465–81, doi:10.1146/annurev-phyto-072910-095423.

Charles Darwin's personal library: I. G. Loskutov, *Vavilov and His Institute: A History of the World Collection of Plant Genetic Resources in Russia* (International Plant Genetic Resources Institute, 1999).

his theory that the greatest genetic diversity: N. I. Vavilov and V. F. Dorofeev, *Origin and Geography of Cultivated Plants* (Cambridge University Press, 1992).

has not stood the test of time: J. Dvorak et al., "NI Vavilov's Theory of Centres of Diversity in the Light of Current Understanding of Wheat Diversity, Domestication and Evolution," *Czech Journal of Genetics and Plant Breeding* 47 (2011): S20–S27.

For twenty years the manuscript was believed lost: S. Reznik and Y. Vavilov, "The Russian Scientist Nicolay Vavilov," in *Five Continents by Nicolay Ivanovich Vavilov*, trans. Doris Löve (IPGRI; VIR, 1997), xvii–xxix.

Vavilov wrote: quoted in G. P. Nabhan, *Where Our Food Comes From: Retracing Nikolay Vavilov's Quest to End Famine* (Island Press Shearwater Books, 2009).

weedy relative from which it was domesticated: A. L. Ingram and J. J. Doyle, "The Origin and Evolution of *Eragrostis tef* (Poaceae) and Related Polyploids: Evidence from Nuclear Waxy and Plastid Rps16," *American Journal of Botany* 90, no. 1 (2003): 116–22.

coda to Vavilov's life: Loskutov, *Vavilov and His Institute*.

the Russland-Sammelcommando: Nabhan, *Where Our Food Comes From*.

G. A. Golubev assessed the impact: Ibid., xxiii, 223.

negatively affected by global warming: J. R. Porter et al., *IPCC Fifth Report*, chapter 7: "Food Security and Food Production Systems" (final draft, 2014).

collecting expedition in Persia: N. I. Vavilov, *Five Continents*.

seeds had also become smaller: J. C. Burger et al., "Rapid Phenotypic Divergence of Feral Rye from Domesticated Cereal Rye," *Weed Science* 55, no. 3 (2007): 204–11, doi:10.1614/WS-06-177.1.

the historian V. Gordon Childe: V. G. Childe, *Man Makes Himself* (Spokesman, 1936).

A study compared the number: G. H. Perry et al., "Diet and the Evolution of Human Amylase Gene Copy Number Variation," *Nature Genetics* 39, no. 10 (2007): 1256–60, doi:10.1038/ng2123.

quite the reverse happened: A. L. Mandel and P. A. S. Breslin, "High Endogenous Salivary Amylase Activity Is Associated with Improved Glycemic Homeostasis Following Starch Ingestion in Adults," *Journal of Nutrition* 142, no. 5 (2012): 853–58, doi:10.3945/jn.111.156984.

digestive system of dogs: E. Axelsson et al., "The Genomic Signature of Dog Domestication Reveals Adaptation to a Starch-Rich Diet," *Nature* 495, no. 7441 (2013): 360–64, doi:10.1038/nature11837.

## CHAPTER FIVE

quite possibly around deep-sea hydrothermal vents: W. Martin et al., "Hydrothermal Vents and the Origin of Life," *Nature Reviews Microbiology* 6, no. 11 (2008): 805–14, doi:10.1038/nrmicro1991; W. F. Martin et al., "Energy at Life's Origin," *Science* 344, no. 6188 (2014): 1092–93, doi:10.1126/science.1251653.

letter written in 1871: C. Darwin, "Letter to J. D. Hooker 1st Feb. 1871," https://www .darwinproject.ac.uk/letter/DCP-LETT-7471.xml (accessed November 5, 2016).

primordial soup: J. B. S. Haldane, "The Origin of Life," *Rationalist Annual* 3 (1929): 3–10.

primordial crêpe or even a primordial vinaigrette: H. S. Bernhardt and W. P. Tate, "Primordial Soup or Vinaigrette: Did the RNA World Evolve at Acidic pH?," *Biology Direct* 7 (2012), doi:10.1186/1745-6150-7-4; G. von Kiedrowski, "Origins of Life—Primordial Soup or Crepes?," *Nature* 381, no. 6577 (1996): 20–21, doi:10.1038/381020a0.

starting with just polysaccharides: V. Tolstoguzov, "Why Are Polysaccharides Necessary?," *Food Hydrocolloids* 18, no. 5 (2004): 873–77, doi:10.1016/j.foodhyd.2003 .11.011.

soup is the basis of our national diet: J. A. Brillat-Savarin, *The Physiology of Taste* (Everyman, 2009), 85.

Beautiful Soup, so rich and green: The Mock Turtle's song from *Alice in Wonderland*.

Harold McGee: H. McGee, *McGee on Food and Cooking* (Hodder & Stoughton, 2004).

sixth taste: R. S. Keast and A. Costanzo, "Is Fat the Sixth Taste Primary? Evidence and Implications," *Flavour* 4, no. 1 (2015): 1–7, doi:10.1186/2044-7248-4-5.

published a paper in Japanese: K. Ikeda, "New Seasonings," *Chemical Senses* 27, no. 9 (2002): 847–49, doi:10.1093/chemse/27.9.847 (translated from the Japanese original published in 1909).

seaweeds from the most saline oceans: O. G. Mouritsen, *Seaweeds: Edible, Available, and Sustainable* (University of Chicago Press, 2013).

trigger an umami taste bomb: O. G. Mouritsen et al., *Umami: Unlocking the Secrets of the Fifth Taste* (Columbia University Press, 2014).

starting point for any soup: L. Bareham, *A Celebration of Soup* (Michael Joseph, 1993).

supplied by inosinate: K. Kurihara, "Glutamate: From Discovery as a Food Flavor to Role as a Basic Taste (Umami)," *American Journal of Clinical Nutrition* 90, no. 3 (2009): 719S–22S, doi:10.3945/ajcn.2009.27462D.

its existence as a fifth taste: B. Lindemann et al., "The Discovery of Umami," *Chemical Senses* 27, no. 9 (2002): 843–44, doi:10.1093/chemse/27.9.843.

tests of the quality of soy sauce: Ikeda, "New Seasonings."

cells in taste buds: N. Chaudhari et al., "A Metabotropic Glutamate Receptor Variant Functions as a Taste Receptor," *Nature Neuroscience* 3, no. 2 (2000): 113–19, doi:10.1038/72053.

the ability to taste sugar: P. H. Jiang et al., "Major Taste Loss in Carnivorous Mammals," *Proceedings of the National Academy of Sciences of the United States of America* 109, no. 13 (2012): 4956–61, doi:10.1073/pnas.1118360109.

Studies in mice: J. Chandrashekar et al., "The Cells and Peripheral Representation of Sodium Taste in Mice," *Nature* 464, no. 7286 (2010): 297–301, doi:10.1038/nature08783.

cucumber beetle: C. P. Da Costa and C. M. Jones, "Cucumber Beetle Resistance and Mite Susceptibility Controlled by the Bitter Gene in *Cucumis sativus* L," *Science* 172, no. 3988 (1971): 1145–46, doi:10.1126/science.172.3988.1145.

thick soups such as cream of onion: R. Man and R. Weir, *The Mustard Book* (Grub Street, 2010).

bitter taste of hops: D. Intelmann et al., "Three TAS2R Bitter Taste Receptors Mediate the Psychophysical Responses to Bitter Compounds of Hops (*Humulus lupulus* L.) and Beer," *Chemosensory Perception* 2, no. 3 (2009): 118–32, doi:10.1007/s12078-009-9049-1.

parted company 93 million years ago: http://www.timetree.org/index.php?taxon_a=mouse&taxon_b=human&submit=Search (accessed October 28, 2014).

receptor genes for bitter compounds: D. Y. Li and J. Z. Zhang, "Diet Shapes the Evolution of the Vertebrate Bitter Taste Receptor Gene Repertoire," *Molecular Biology and Evolution* 31, no. 2 (2014): 303–9, doi:10.1093/molbev/mst219.

The 11 pseudogenes: Y. Go et al., "Lineage-Specific Loss of Function of Bitter Taste Receptor Genes in Humans and Nonhuman Primates," *Genetics* 170, no. 1 (2005): 313–26, doi:10.1534/genetics.104.037523.

Mice engineered: K. L. Mueller et al., "The Receptors and Coding Logic for Bitter Taste," *Nature* 434, no. 7030 (2005): 221–25, doi:10.1038/nature03366.

react differently to sour-tasting things: D. G. Liem and J. A. Mennella, "Heightened Sour Preferences during Childhood," *Chemical Senses* 28, no. 2 (2003): 173–80.

ability to taste: D. Drayna, "Human Taste Genetics," *Annual Review of Genomics and Human Genetics* 6 (2005): 217–35.

Edinburgh Zoo: R. A. Fisher et al., "Taste-Testing the Anthropoid Apes," *Nature* 144 (1939): 750.

evenly balanced: Drayna, "Human Taste Genetics."

anti-cancer properties: Y. Shang et al., "Biosynthesis, Regulation, and Domestication of Bitterness in Cucumber," *Science* 346, no. 6213 (2014): 1084–88, doi:10.1126/science.1259215.

## CHAPTER SIX

can then be enjoyed as sashimi: O. G. Mouritsen et al., *Umami: Unlocking the Secrets of the Fifth Taste* (Columbia University Press, 2014).

Even pain receptors: F. Viana, "Chemosensory Properties of the Trigeminal System," *ACS Chemical Neuroscience* 2, no. 1 (2011): 38–50, doi:10.1021/cn100102c.

Père Polycarpe Poncelet: "*Chimie du goût et de l'odorat* [1st ed., 1755]," described by A. Davidson, "Tastes, Aromas, Flavours," in *Oxford Symposium on Food and Cookery, 1987: Taste*, ed. T. Jaine (Prospect Books, 1988): 9–14.

Aristotle: quoted in G. M. Shepherd, *Neurogastronomy: How the Brain Creates Flavor and Why It Matters* (Columbia University Press, 2012), 12.

different olfactory receptors: Y. Niimura, "Olfactory Receptor Multigene Family in

Vertebrates: From the Viewpoint of Evolutionary Genomics," *Current Genomics* 13, no. 2 (2012): 103–14.

African elephants: Y. Niimura et al., "Extreme Expansion of the Olfactory Receptor Gene Repertoire in African Elephants and Evolutionary Dynamics of Orthologous Gene Groups in 13 Placental Mammals," *Genome Research* 24, no. 9 (2014): 1485–96, doi:10.1101/gr.169532.113.

a great deal of evolutionary change: Y. Niimura and M. Nei, "Extensive Gains and Losses of Olfactory Receptor Genes in Mammalian Evolution," *PLOS ONE* 2, no. 8 (2007), doi:10.1371/journal.pone.0000708.

more than a trillion distinct smells: C. Bushdid et al., "Humans Can Discriminate More than 1 Trillion Olfactory Stimuli," *Science* 343, no. 6177 (2014): 1370–72, doi:10.1126/science.1249168.

combine them in all the ways imaginable: M. Auvray and C. Spence, "The Multisensory Perception of Flavor," *Consciousness and Cognition* 17, no. 3 (2008): 1016–31, doi:10.1016/j.concog.2007.06.005.

we remain unaware: G. M. Shepherd, "The Human Sense of Smell: Are We Better than We Think?," *PLOS Biology* 2, no. 5 (2004): e146, doi:10.1371/journal.pbio.0020146.

600 alleles each: T. Olender et al., "Personal Receptor Repertoires: Olfaction as a Model," *BMC Genomics* 13 (2012), doi:10.1186/1471-2164-13-414.

Every one of these alleles is used: B. Keverne, "Monoallelic Gene Expression and Mammalian Evolution," *Bioessays* 31, no. 12 (2009): 1318–26, doi:10.1002/bies .200900074.

liked cilantro or not: N. Eriksson et al., "A Genetic Variant Near Olfactory Receptor Genes Influences Cilantro Preference," *Flavour* 1, no. 22 (2012), doi:10.1186/2044 -7248-1-22.

how evolution has adapted their muscles: H. McGee, *McGee on Food and Cooking* (Hodder & Stoughton, 2004).

the use of garum: R. I. Curtis, "Umami and the Foods of Classical Antiquity," *American Journal of Clinical Nutrition* 90, no. 3 (2009): 712S–18S, doi:10.3945/ajcn.2009 .27462C.

only large-scale factory industry: A. Dalby and S. Grainger, *The Classical Cookbook* (British Museum Press, 1996).

garum tycoon from ill-fated Pompeii: Curtis, "Umami and the Foods of Classical Antiquity."

## CHAPTER SEVEN

Meat also contributes: N. Mann, "Dietary Lean Red Meat and Human Evolution," *European Journal of Nutrition* 39, no. 2 (2000): 71–79, doi:10.1007/s003940050005.

association with these parasites: E. P. Hoberg et al., "Out of Africa: Origins of the *Taenia* Tapeworms in Humans," *Proceedings of the Royal Society of London: Series B, Biological Sciences* 268, no. 1469 (2001): 781–87.

Trichinella spiralis: D. S. Zarlenga et al., "Post-Miocene Expansion, Colonization, and Host Switching Drove Speciation among Extant Nematodes of the Archaic Genus *Trichinella*," *Proceedings of the National Academy of Sciences of the United States of America* 103, no. 19 (2006): 7354–59, doi:10.1073/pnas.0602466103.

well-protected against heat shock: G. H. Perry, "Parasites and Human Evolution," *Evolutionary Anthropology* 23, no. 6 (2014): 218–28, doi:10.1002/evan.21427.

first recognizable animal: M. Aubert et al., "Pleistocene Cave Art from Sulawesi, Indonesia," *Nature* 514, no. 7521 (2014): 223–27, doi:10.1038/nature13422.

*Babyrousa babyrussa*: L. Watson, *The Whole Hog: Exploring the Extraordinary Potential of Pigs* (Profile, 2004).

Chauvet cave: http://www.bradshawfoundation.com/chauvet/ (accessed July 14, 2015);
J. Combier and G. Jouve, "Nouvelles recherches sur l'identité culturelle et stylistique
de la grotte Chauvet et sur sa datation par la méthode du 14C," *L'Anthropologie* 118,
no. 2 (2014): 115–51, doi:10.1016/j.anthro.2013.12.001.

reindeer meat: S. Gaudzinski-Windheuser and L. Niven, "Hominin Subsistence Patterns during the Middle and Late Paleolithic in Northwestern Europe," in *The Evolution of Hominin Diets*, Vertebrate Paleobiology and Paleoanthropology, ed. J.-J.
Hublin and M. Richards (Springer Netherlands, 2009), 99–111.

a cobblestone: M. Mariotti Lippi et al., "Multistep Food Plant Processing at Grotta
Paglicci (Southern Italy) around 32,600 Cal B.P.," *Proceedings of the National Academy
of Sciences of the United States of America* 112, no. 39 (2015): 12075–80, doi:10.1073
/pnas.1505213112.

caches that wild field mice make: M. Jones, "Moving North: Archaeobotanical Evidence
for Plant Diet in Middle and Upper Paleolithic Europe," in *The Evolution of Hominin
Diets*, Vertebrate Paleobiology and Paleoanthropology, ed. J.-J. Hublin and M. Richards (Springer Netherlands, 2009), 171–80.

vegetation changed: E. Willerslev et al., "Fifty Thousand Years of Arctic Vegetation
and Megafaunal Diet," *Nature* 506, no. 7486 (2014): 47–51, doi:10.1038/nature
12921.

variety of gray wolf: J. A. Leonard et al., "Megafaunal Extinctions and the Disappearance of a Specialized Wolf Ecomorph," *Current Biology* 17, no. 13 (2007): 1146–50,
doi:10.1016/j.cub.2007.05.072.

remnant populations: M. Hofreiter and I. Barnes, "Diversity Lost: Are All Holarc-
tic Large Mammal Species Just Relict Populations?," *BMC Biology* 8 (2010): 46,
doi:10.1186/1741-7007-8-46.

hunters finishing off: A. J. Stuart, "Late Quaternary Megafaunal Extinctions on the Continents: A Short Review," *Geological Journal* 50, no. 3 (2015): 338–63, doi:10.1002
/gj.2633.

favorite food: H. Bocherens et al., "Reconstruction of the Gravettian Food-Web at
Předmostí I Using Multi-Isotopic Tracking (13C, 15N, 34S) of Bone Collagen," *Quaternary International* 359 (2015): 211–28, doi:10.1016/j.quaint.2014.09.044.

mammoth was an evergreen favorite: P. Shipman, "How Do You Kill 86 Mammoths?
Taphonomic Investigations of Mammoth Megasites," *Quaternary International* 359–
60 (2015): 38–46, doi:10.1016/j.quaint.2014.04.048.

Wrangel Island: A. J. Stuart et al., "Pleistocene to Holocene Extinction Dynamics in
Giant Deer and Woolly Mammoth," *Nature* 431 (2004): 684–89.

began to broaden their diet: M. C. Stiner and N. D. Munro, "Approaches to Prehistoric
Diet Breadth, Demography, and Prey Ranking Systems in Time and Space," *Journal
of Archaeological Method and Theory* 9, no. 2 (June 2002): 181–214.

Ohalo II: L. A. Maher et al., "The Pre-Natufian Epipaleolithic: Long-Term Behavioral
Trends in the Levant," *Evolutionary Anthropology* 21, no. 2 (2012): 69–81, doi:10.1002
/evan.21307.

weeds of cultivation: A. Snir et al., "The Origin of Cultivation and Proto-Weeds,
Long Before Neolithic Farming," *PLOS ONE* 10, no. 7 (2015), doi:10.1371/journal
.pone.0131422.

people at Ohalo II ate: D. Nadel et al., "On the Shore of a Fluctuating Lake: Environmental Evidence from Ohalo II (19,500 BP)," *Israel Journal of Earth Sciences* 53, nos. 3–4,
special issue (2004): 207–23, doi:10.1560/v3cu-ebr7-ukat-uca6.

site near Haifa called el-Wad: R. Yeshurun et al., "Intensification and Sedentism in
the Terminal Pleistocene Natufian Sequence of el-Wad Terrace (Israel)," *Journal of
Human Evolution* 70 (2014): 16–35, doi:10.1016/j.jhevol.2014.02.011.

Aşıklı Höyük: M. C. Stiner et al., "A Forager-Herder Trade-Off, from Broad-Spectrum
Hunting to Sheep Management at A ıklı Höyük, Turkey," *Proceedings of the National*

*Academy of Sciences of the United States of America* 111, no. 23 (2014): 8404–9, doi:10.1073/pnas.1322723111.

children born per woman nearly doubled: E. Guerrero, S. Naji, and J.-P. Bocquet-Appel, "The Signal of the Neolithic Demographic Transition in the Levant," in *The Neolithic Demographic Transition and Its Consequences*, ed. J.-P. Bocquet-Appel and O. Bar-Yosef (Springer, 2008), 57–80, doi:10.1007/978-1-4020-8539-0_4.

global phenomenon: P. Bellwood and M. Oxenham, "The Expansions of Farming Societies and the Role of the Neolithic Demographic Transition," in ibid., 13–34, doi:10.1007/978-1-4020-8539-0_2.

You'll get mixed up: Dr. Seuss, *Oh, the Places You'll Go!* (Random House, 1990).

genetic affinity with modern domestic chickens: H. Xiang et al., "Early Holocene Chicken Domestication in Northern China," *Proceedings of the National Academy of Sciences of the United States of America* 111, no. 49 (2014): 17564–69, doi:10.1073/pnas.1411882111.

independently domesticated: S. Kanginakudru et al., "Genetic Evidence from Indian Red Jungle Fowl Corroborates Multiple Domestication of Modern Day Chicken," *BMC Evolutionary Biology* 8 (2008): 174, doi:10.1186/1471-2148-8-174; Y. P. Liu et al., "Multiple Maternal Origins of Chickens: Out of the Asian Jungles," *Molecular Phylogenetics and Evolution* 38, no. 1 (2006): 12–19, doi:10.1016/j.ympev.2005.09.014.

an edible souvenir: A. A. Storey et al., "Investigating the Global Dispersal of Chickens in Prehistory Using Ancient Mitochondrial DNA Signatures," *PLOS ONE* 7, no. 7 (2012), doi:10.1371/journal.pone.0039171.

the grey jungle fowl: J. Eriksson et al., "Identification of the Yellow Skin Gene Reveals a Hybrid Origin of the Domestic Chicken," *PLOS Genetics* 4, no. 2 (2008), doi:10.1371/journal.pgen.1000010.

three separate introductions: J. M. Mwacharo et al., "The History of African Village Chickens: An Archaeological and Molecular Perspective," *African Archaeological Review* 30, no. 1 (2013): 97–114, doi:10.1007/s10437-013-9128-1; J. M. Mwacharo et al., "Reconstructing the Origin and Dispersal Patterns of Village Chickens across East Africa: Insights from Autosomal Markers," *Molecular Ecology* 22, no. 10 (2013): 2683–97, doi:10.1111/mec.12294.

most heroic of all migrations: P. V. Kirch, "Peopling of the Pacific: A Holistic Anthropological Perspective," *Annual Review of Anthropology* 39, no. 1 (2010): 131–48, doi:10.1146/annurev.anthro.012809.104936; J. M. Wilmshurst et al., "High-Precision Radiocarbon Dating Shows Recent and Rapid Initial Human Colonization of East Polynesia," *Proceedings of the National Academy of Sciences of the United States of America* 108, no. 5 (2011): 1815–20, doi:10.1073/pnas.1015876108.

chicken houses: J. Diamond, *Collapse: How Societies Choose to Fail or Survive* (Allen Lane, 2005).

prehistoric Polynesian chickens: Storey et al., "Investigating the Global Dispersal of Chickens"; A. A. Storey, "Polynesian Chickens in the New World: A Detailed Application of a Commensal Approach," *Archaeology in Oceania* 48 (2013): 101–19, doi:10.1002/arco.5007.

Spanish conquistador Francisco Pizarro: S. M. Fitzpatrick and R. Callaghan, "Examining Dispersal Mechanisms for the Translocation of Chicken (*Gallus gallus*) from Polynesia to South America," *Journal of Archaeological Science* 36, no. 2 (2009): 214–23, doi:10.1016/j.jas.2008.09.002.

ball bearings: J. Flenley and P. Bahn, *The Enigmas of Easter Island* (Oxford University Press, 2002).

Ecuador and Peru: C. Roullier et al., "Historical Collections Reveal Patterns of Diffusion of Sweet Potato in Oceania Obscured by Modern Plant Movements and Recombination," *Proceedings of the National Academy of Sciences of the United States of America* 110, no. 6 (2013): 2205–10, doi:10.1073/pnas.1211049110.

were made welcome: J. V. Moreno-Mayar et al., "Genome-Wide Ancestry Patterns in Rapanui Suggest Pre-European Admixture with Native Americans," *Current Biology* 24, no. 21 (2014): 2518–25, doi:10.1016/j.cub.2014.09.057.

15,000 years ago: D. F. Morey, "In Search of Paleolithic Dogs: A Quest with Mixed Results," *Journal of Archaeological Science* 52 (2014): 300–307, doi:10.1016/j.jas.2014.08.015.

twice as old: Shipman, "How Do You Kill 86 Mammoths?"

all points of the compass: F. H. Lv et al., "Mitogenomic Meta-Analysis Identifies Two Phases of Migration in the History of Eastern Eurasian Sheep," *Molecular Biology and Evolution* 32, no. 10 (2015): 2515–33, doi:10.1093/molbev/msv139.

northern China: J. Dodson et al., "Oldest Directly Dated Remains of Sheep in China," *Scientific Reports* 4 (2014), doi:10.1038/srep07170.

1,500 different breeds: P. Taberlet et al., "Conservation Genetics of Cattle, Sheep, and Goats," *Comptes Rendus Biologies* 334, no. 3 (2011): 247–54, doi:10.1016/j.crvi.2010.12.007.

fat tail: M. H. Moradi et al., "Genomic Scan of Selective Sweeps in Thin and Fat Tail Sheep Breeds for Identifying of Candidate Regions Associated with Fat Deposition," *BMC Genetics* 13 (2012): 10, doi:10.1186/1471-2156-13-10.

traditional cooking medium: J. Tilsley-Benham, "Sheep with Two Tails: Sheep's Tail Fat as a Cooking Medium in the Middle East," *Oxford Symposium on Food & Cookery, 1986: The Cooking Medium: Proceedings*, ed. T. Jaine (Prospect Books, 1987), 46–50.

the transition that took place: N. Marom and G. Bar-Oz, "The Prey Pathway: A Regional History of Cattle (*Bos taurus*) and Pig (*Sus scrofa*) Domestication in the Northern Jordan Valley, Israel," *PLOS ONE* 8, no. 2 (2013): e55958, doi:10.1371/journal.pone.0055958.

domesticated three times: J. E. Decker et al., "Worldwide Patterns of Ancestry, Divergence, and Admixture in Domesticated Cattle," *PLOS Genetics* 10, no. 3 (2014), doi:10.1371/journal.pgen.1004254.

Studies of human genetics: W. Haak et al., "Ancient DNA from European Early Neolithic Farmers Reveals Their Near Eastern Affinities," *PLOS Biology* 8, no. 11 (2010): e1000536, doi:10.1371/journal.pbio.1000536; Q. M. Fu et al., "Complete Mitochondrial Genomes Reveal Neolithic Expansion into Europe," *PLOS ONE* 7, no. 3 (2012), doi:10.1371/journal.pone.0032473.

dispersed into Europe: R. Pinhasi et al., "Tracing the Origin and Spread of Agriculture in Europe," *PLOS Biology* 3, no. 12 (2005): e410, doi:10.1371/journal.pbio.0030410.

first farmers in Anatolia: A. Gibbons, "First Farmers' Motley Roots," *Science* 353, no. 6296 (2016): 207–8.

nomadic pastoralism: O. Hanotte et al., "African Pastoralism: Genetic Imprints of Origins and Migrations," *Science* 296, no. 5566 (2002): 336–39, doi:10.1126/science.1069878.

evolutionary home: L. A. F. Frantz et al., "Genome Sequencing Reveals Fine Scale Diversification and Reticulation History during Speciation in Sus," *Genome Biology* 14, no. 9 (2013), doi:10.1186/gb-2013-14-9-r107.

six or seven times: G. Larson et al., "Worldwide Phylogeography of Wild Boar Reveals Multiple Centers of Pig Domestication," *Science* 307, no. 5715 (2005): 1618–21.

twice in China: G. S. Wu et al., "Population Phylogenomic Analysis of Mitochondrial DNA in Wild Boars and Domestic Pigs Revealed Multiple Domestication Events in East Asia," *Genome Biology* 8, no. 11 (2007), doi:10.1186/gb-2007-8-11-r245.

Vietnam: G. Larson et al., "Phylogeny and Ancient DNA of *Sus* Provides Insights into Neolithic Expansion in Island Southeast Asia and Oceania," *Proceedings of the National Academy of Sciences of the United States of America* 104, no. 12 (2007): 4834–39, doi:10.1073/pnas.0607753104.

religious taboo: Watson, *The Whole Hog.*

red deer: J. Clutton-Brock, *A Natural History of Domesticated Mammals* (Cambridge University Press, 1999).

domesticated twice: K. H. Roed et al., "Genetic Analyses Reveal Independent Domestication Origins of Eurasian Reindeer," *Proceedings of the Royal Society B: Biological Sciences* 275, no. 1645 (2008): 1849–55, doi:10.1098/rspb.2008.0332.

*The Variation of Animals and Plants under Domestication*: C. Darwin, *The Variation of Animals and Plants under Domestication* (John Murray, 1868).

Douglas Adams: D. Adams, *The Restaurant at the End of the Universe* (Random House, 2008).

Siberian silver foxes: L. Trut et al., "Animal Evolution during Domestication: The Domesticated Fox as a Model," *Bioessays* 31, no. 3 (2009): 349–60, doi:10.1002/bies.200800070.

Russian scientists: Ibid.

no one has yet been able to find it: G. Larson and D. Q. Fuller, "The Evolution of Animal Domestication," *Annual Review of Ecology, Evolution, and Systematics* 45, no. 1 (2014): 115–36, doi:10.1146/annurev-ecolsys-110512-135813.

another explanation: A. S. Wilkins et al., "The 'Domestication Syndrome' in Mammals: A Unified Explanation Based on Neural Crest Cell Behavior and Genetics," *Genetics* 197, no. 3 (2014): 795–808, doi:10.1534/genetics.114.165423.

hunter-gatherers who live this way today: A. Strohle and A. Hahn, "Diets of Modern Hunter-Gatherers Vary Substantially in Their Carbohydrate Content Depending on Ecoenvironments: Results from an Ethnographic Analysis," *Nutrition Research* 31, no. 6 (2011): 429–35, doi:10.1016/j.nutres.2011.05.003; C. Higham, "Hunter-Gatherers in Southeast Asia: From Prehistory to the Present," *Human Biology* 85, no. 1–3 (2013): 21–43.

## CHAPTER EIGHT

4,000 different species: S. Proches et al., "Plant Diversity in the Human Diet: Weak Phylogenetic Signal Indicates Breadth," *Bioscience* 58, no. 2 (2008): 151–59, doi:10.1641/b580209.

lectins that in nature protect beans: G. Vandenborre et al., "Plant Lectins as Defense Proteins against Phytophagous Insects," *Phytochemistry* 72, no. 13 (2011): 1538–50, doi:10.1016/j.phytochem.2011.02.024.

poisonous results: J. C. Rodhouse et al., "Red Kidney Bean Poisoning in the UK—An Analysis of 50 Suspected Incidents between 1976 and 1989," *Epidemiology and Infection* 105, no. 3 (1990): 485–91.

walking stick: http://jerseyeveningpost.com/island-life/history-heritage/giant-cabbage/ (accessed April 28, 2015).

Trophy: L. H. Bailey, *The Survival of the Unlike: A Collection of Evolution Essays Suggested by the Study of Domestic Plants* (Macmillan, 1897).

remarkable changes made by artificial selection: Y. Bai and P. Lindhout, "Domestication and Breeding of Tomatoes: What Have We Gained and What Can We Gain in the Future?," *Annals of Botany* 100, no. 5 (2007): 1085–94, doi:10.1093/aob/mcm150.

a handful of genes: E. van der Knaap et al., "What Lies beyond the Eye: The Molecular Mechanisms Regulating Tomato Fruit Weight and Shape," *Frontiers in Plant Science* 5 (2014), doi:10.3389/fpls.2014.00227.

increased its size: Bailey, *The Survival of the Unlike*, 485.

big changes in crops: J. F. Hancock, *Plant Evolution and the Origin of Crop Species* (CABI, 2012).

domesticated by the Maya: J. A. Jenkins, "The Origin of the Cultivated Tomato," *Economic Botany* 2, no. 4 (1948): 379–92, doi:10.1007/BF02859492.

Lilliputian fruits: Hancock, *Plant Evolution and the Origin of Crop Species.*

a huge variety of *tomatl*: S. D. Coe, *America's First Cuisines* (University of Texas Press, 1994).

heirloom tomato website: http://www.heirloomtomatoes.net/Varieties.html (accessed April 16, 2015).

70 crops in cultivation: O. F. Cook, "Peru as a Center of Domestication: Tracing the Origin of Civilization through Domesticated Plants (continued)," *Journal of Heredity* 16, no. 3 (1925): 95–110.

16,000–17,000 years ago: N. Misarti et al., "Early Retreat of the Alaska Peninsula Glacier Complex and the Implications for Coastal Migrations of First Americans," *Quaternary Science Reviews* 48 (2012): 1–6, doi:10.1016/j.quascirev.2012.05.014.

then-established view: T. D. Dillehay, "Battle of Monte Verde," *The Sciences* (January/February 1997): 28–33.

including wild potatoes: T. D. Dillehay et al., "Monte Verde: Seaweed, Food, Medicine, and the Peopling of South America," *Science* 320, no. 5877 (2008): 784–86, doi:10.1126/science.1156533.

eating peanuts, squash: D. R. Piperno and T. D. Dillehay, "Starch Grains on Human Teeth Reveal Early Broad Crop Diet in Northern Peru," *Proceedings of the National Academy of Sciences of the United States of America* 105, no. 50 (2008): 19622–27, doi:10.1073/pnas.0808752105.

plant remains: T. D. Dillehay et al., "Preceramic Adoption of Peanut, Squash, and Cotton in Northern Peru," *Science* 316, no. 5833 (2007): 1890–93, doi:10.1126/science.1141395.

a single wild species called *Solanum candolleanum*: D. M. Spooner et al., "Systematics, Diversity, Genetics, and Evolution of Wild and Cultivated Potatoes," *Botanical Review* 80, no. 4 (2014): 283–383, doi:10.1007/s12229-014-9146-y.

3,000 potato landraces: Ibid.

*Solanum hydrothericum*: National Research Council, *Lost Crops of the Incas: Little-Known Plants of the Andes with Promise for Worldwide Cultivation* (National Academy Press, 1989).

resist aphids better: K. L. Flanders et al., "Insect Resistance in Potatoes—Sources, Evolutionary Relationships, Morphological and Chemical Defenses, and Ecogeographical Associations," *Euphytica* 61, no. 2 (1992): 83–111, doi:10.1007/bf00026800.

resistance to late blight: G. M. Rauscher et al., "Characterization and Mapping of $R_{Pi\text{-}ber}$, a Novel Potato Late Blight Resistance Gene from *Solanum berthaultii*," *Theoretical and Applied Genetics* 112, no. 4 (2006): 674–87, doi:10.1007/s00122-005-0171-4.

a million people died: J. Reader, *The Untold History of the Potato* (Vintage, 2009).

evolved resistance: Y. T. Hwang et al., "Evolution and Management of the Irish Potato Famine Pathogen *Phytophthora infestans* in Canada and the United States," *American Journal of Potato Research* 91, no. 6 (2014): 579–93, doi:10.1007/s12230-014-9401-0.

*chuño*: Reader, *The Untold History of the Potato.*

garden dedicated to the sun: Ibid.

move thousands of people: National Research Council, *Lost Crops of the Incas.*

almost 20 other root crops: Ibid.

Manioc (*Manihot esculenta*): K. M. Olsen and B. A. Schaal, "Evidence on the Origin of Cassava: Phylogeography of *Manihot esculenta*," *Proceedings of the National Academy of Sciences of the United States of America* 96, no. 10 (1999): 5586–91, doi:10.1073/pnas.96.10.5586.

grown in gardens by forest dwellers: M. Arroyo-Kalin, "The Amazonian Formative: Crop Domestication and Anthropogenic Soils," *Diversity* 2, no. 4 (2010): 473–504, doi:10.3390/d2040473.

non-toxic varieties of manioc: D. McKey et al., "Chemical Ecology in Coupled Human and Natural Systems: People, Manioc, Multitrophic Interactions and Global Change," *Chemoecology* 20, no. 2 (2010): 109–33, doi:10.1007/s00049-010-0047-1.

300 million years ago: C. C. Labandeira, "Early History of Arthropod and Vascular Plant Associations," *Annual Review of Earth and Planetary Sciences* 26 (1998): 329–77, doi:10.1146/annurev.earth.26.1.329.

similar to the one that produces cyanogenic glycosides: J. E. Rodman et al., "Parallel Evolution of Glucosinolate Biosynthesis Inferred from Congruent Nuclear and Plastid Gene Phylogenies," *American Journal of Botany* 85, no. 7 (1998): 997–1006, doi:10.2307/2446366.

tumor-suppressing effects: M. Traka and R. Mithen, "Glucosinolates, Isothiocyanates and Human Health," *Phytochemistry Reviews* 8, no. 1 (2009): 269–82, doi:10.1007/s11101-008-9103-7.

detoxification mechanism evolved: C. W. Wheat et al., "The Genetic Basis of a Plant-Insect Coevolutionary Key Innovation," *Proceedings of the National Academy of Sciences of the United States of America* 104, no. 51 (2007): 20427–31, doi:10.1073/pnas.0706229104.

a thousand new butterfly species: M. F. Braby and J. W. H. Trueman, "Evolution of Larval Host Plant Associations and Adaptive Radiation in Pierid Butterflies," *Journal of Evolutionary Biology* 19, no. 5 (2006): 1677–90.

tolerate cyanide: E. J. Stauber et al., "Turning the 'Mustard Oil Bomb' into a 'Cyanide Bomb': Aromatic Glucosinolate Metabolism in a Specialist Insect Herbivore," *PLOS ONE* 7, no. 4 (2012), doi:10.1371/journal.pone.0035545.

experimental results: T. Zust et al., "Natural Enemies Drive Geographic Variation in Plant Defenses," *Science* 338, no. 6103 (2012): 116–19, doi:10.1126/science.1226397.

greatest genetic variation: B. Pujol et al., "Microevolution in Agricultural Environments: How a Traditional Amerindian Farming Practice Favors Heterozygosity in Cassava (*Manihot esculenta* Crantz, Euphorbiaceae)," *Ecology Letters* 8, no. 2 (2005): 138–47, doi:10.1111/j.1461-0248.2004.00708.x.

drew a diagram: I. Ahuja et al., "Defence Mechanisms of Brassicaceae: Implications for Plant-Insect Interactions and Potential for Integrated Pest Management: A Review," *Agronomy for Sustainable Development* 30, no. 2 (2010): 311–48, doi:10.1051/agro/2009025.

Modern genomic analysis: T. Arias et al., "Diversification Times among Brassica (Brassicaceae) Crops Suggest Hybrid Formation after 20 Million Years of Divergence," *American Journal of Botany* 101, no. 1 (2014): 86–91, doi:10.3732/ajb.1300312.

a cross between wild black mustard weeds: Hancock, *Plant Evolution and the Origin of Crop Species*.

## CHAPTER NINE

The Arabians say: J. Keay, *The Spice Route: A History* (John Murray, 2005).

Hernán Cortés: J. Turner, *Spice: The History of a Temptation* (Harper Perennial, 2005), 11.

Pharaoh Ramses II: A. Gilboa and D. Namdar, "On the Beginnings of South Asian Spice Trade with the Mediterranean Region: A Review," *Radiocarbon* 57, no. 2 (2015): 265–83, doi:10.2458/azu_rc.57.18562.

The black pepper vine: D. Q. Fuller et al., "Across the Indian Ocean: The Prehistoric Movement of Plants and Animals," *Antiquity* 85, no. 328 (2011): 544–58.

trail of lost Roman coins: Keay, *The Spice Route*.

the Phoenicians: Gilboa and Namdar, "On the Beginnings of South Asian Spice Trade."

some have argued: P. W. Sherman and J. Billing, "Darwinian Gastronomy: Why We Use Spices," *Bioscience* 49, no. 6 (1999): 453–63, doi:10.2307/1313553.

make matters worse: Keay, *The Spice Route*.

*Allium*: E. Block, *Garlic and Other Alliums: The Lore and the Science* (Royal Society of Chemistry Publications, 2010).

half a million carbon atoms: N. Theis and M. Lerdau, "The Evolution of Function in Plant Secondary Metabolites," *International Journal of Plant Sciences* 164, no. 3 (May 2003): S93–S102.

two phases of construction: R. Firn, *Nature's Chemicals: The Natural Products That Shaped Our World* (Oxford University Press, 2010).

40,000 chemical products of the terpenoid pathway: S. Steiger et al., "The Origin and Dynamic Evolution of Chemical Information Transfer," *Proceedings of the Royal Society of London: Series B, Biological Sciences* 278, no. 1708 (2011): 970–79, doi:10.1098 /rspb.2010.2285.

mixtures of monoterpenes: Firn, *Nature's Chemicals*.

wild thyme: J. D. Thompson, *Plant Evolution in the Mediterranean* (Oxford University Press, 2005).

The explanation: J. Thompson et al., "Evolution of a Genetic Polymorphism with Climate Change in a Mediterranean Landscape," *Proceedings of the National Academy of Sciences of the United States of America* 110, no. 8 (2013): 2893–97, doi:10.1073 /pnas.1215833110; J. D. Thompson et al., "Ongoing Adaptation to Mediterranean Climate Extremes in a Chemically Polymorphic Plant," *Ecological Monographs* 77, no. 3 (2007): 421–39, doi:10.1890/06-1973.1.

Rosemary: Thompson, *Plant Evolution in the Mediterranean*.

spices also stimulate pain sensors: D. Julius, "TRP Channels and Pain," *Annual Review of Cell and Developmental Biology* 29 (2013): 355–84, doi:10.1146/annurev-cellbio -101011-155833.

Each TRP type is activated: F. Viana, "Chemosensory Properties of the Trigeminal System," *ACS Chemical Neuroscience* 2, no. 1 (2011): 38–50, doi:10.1021/cn100102c.

Cinnamon only stimulates TRPA1: Ibid.

tarantula venom: J. Siemens et al., "Spider Toxins Activate the Capsaicin Receptor to Produce Inflammatory Pain," *Nature* 444, no. 7116 (2006): 208–12, doi:10.1038 /nature05285.

TRP receptors: S. F. Pedersen et al., "TRP Channels: An Overview," *Cell Calcium* 38, nos. 3–4 (2005): 233–52, doi:10.1016/j.ceca.2005.06.028.

learn to enjoy the stimulation: E. Carstens et al., "It Hurts So Good: Oral Irritation by Spices and Carbonated Drinks and the Underlying Neural Mechanisms," *Food Quality and Preference* 13, nos. 7–8 (October–December 2002): 431–43.

Certain TRP genes: S. Saito and M. Tominaga, "Functional Diversity and Evolutionary Dynamics of ThermoTRP Channels," *Cell Calcium* 57, no. 3 (2015): 214–21, doi:10.1016/j.ceca.2014.12.001.

insensitive to this chemical in birds: S. E. Jordt and D. Julius, "Molecular Basis for Species-Specific Sensitivity to 'Hot' Chili Peppers," *Cell* 108, no. 3 (2002): 421–30, doi:10.1016/s0092-8674(02)00637-2.

Experiments with wild chili: J. J. Tewksbury and G. P. Nabhan, "Seed Dispersal—Directed Deterrence by Capsaicin in Chillies," *Nature* 412, no. 6845 (2001): 403–4.

a single gene called *Pun1*: C. Stewart et al., "Genetic Control of Pungency in *C. chinense* via the *Pun1* Locus," *Journal of Experimental Botany* 58, no. 5 (2007): 979–91, doi:10.1093/jxb/erl243.

a fungus called *Fusarium*: J. J. Tewksbury et al., "Evolutionary Ecology of Pungency in Wild Chilies," *Proceedings of the National Academy of Sciences of the United States of America* 105, no. 33 (2008): 11808–11, doi:10.1073/pnas.0802691105.

half the number of seeds: D. C. Haak et al., "Why Are Not All Chilies Hot? A Trade-Off Limits Pungency," *Proceedings of the Royal Society of London: Series B, Biological Sciences* 279, no. 1735 (2012): 2012–17, doi:10.1098/rspb.2011.2091.

CHAPTER TEN

Fred Plotkin: As told to a master class on opera and food at the Royal Opera House, Covent Garden and broadcast on BBC Radio 4 Food Programme, July 13, 2014, http://www.bbc.co.uk/programmes/b0495lm1 (accessed March 12, 2014).

domesticated in New Guinea: P. H. Moore et al., "Sugarcane: The Crop, the Plant, and Domestication," in *Sugarcane: Physiology, Biochemistry, and Functional Biology* (John Wiley & Sons, 2013), 1–17.

great ape cousins: A. N. Crittenden, "The Importance of Honey Consumption in Human Evolution," *Food and Foodways* 19, no. 4 (2011): 257–73, doi:10.1080/07409710.2011.630618.

consumption of honey by the Hadza: F. W. Marlowe et al., "Honey, Hadza, Hunter-Gatherers, and Human Evolution," *Journal of Human Evolution* 71 (2014): 119–28, doi:10.1016/j.jhevol.2014.03.006.

honeyguides and people do communicate: H. A. Isack and H.-U. Reyer, "Honeyguides and Honey Gatherers: Interspecific Communication in a Symbiotic Relationship," *Science* 243, no. 4896 (1989): 1343–46, doi:10.1126/science.243.4896.1343.

less than a fifth of the time: B. M. Wood et al., "Mutualism and Manipulation in Hadza-Honeyguide Interactions," *Evolution and Human Behavior* 35, no. 6 (2014): 540–46, doi:10.1016/j.evolhumbehav.2014.07.007.

herbs to pacify bees: T. S. Kraft and V. V. Venkataraman, "Could Plant Extracts Have Enabled Hominins to Acquire Honey before the Control of Fire?," *Journal of Human Evolution* 85 (2015): 65–74, doi:10.1016/j.jhevol.2015.05.010.

Pliny the Elder: A. Mayor, "Mad Honey!," *Archaeology* 48, no. 6 (1995): 32–40, doi:10.2307/41771162.

poisoning by mad honey: A. Demircan et al., "Mad Honey Sex: Therapeutic Misadventures from an Ancient Biological Weapon," *Annals of Emergency Medicine* 54, no. 6 (2009): 824–29, http://dx.doi.org/10.1016/j.annemergmed.2009.06.010.

two-thirds of the population: C. L. Ogden et al., "Prevalence of Childhood and Adult Obesity in the United States (2011–2012)," *JAMA* 311, no. 8 (2014): 806–14, doi:10.1001/jama.2014.732.

average across western Europe: M. Ng et al., "Global, Regional, and National Prevalence of Overweight and Obesity in Children and Adults during 1980–2013: A Systematic Analysis for the Global Burden of Disease Study 2013," *The Lancet* 384, no. 9945 (2014): 766–81, doi:10.1016/s0140-6736(14)60460-8.

Hunger has not gone away: A. Sonntag et al. *2014 Global Hunger Index: The Challenge of Hidden Hunger* (International Food Policy Research Institute, 2014).

James Neel: J. V. Neel, "Diabetes Mellitus—a Thrifty Genotype Rendered Detrimental by Progress," *American Journal of Human Genetics* 14, no. 4 (1962): 353–57.

frequency of famines: J. C. Berbesque et al., "Hunter-Gatherers Have Less Famine than Agriculturalists," *Biology Letters* 10, no. 1 (2014), doi:10.1098/rsbl.2013.0853.

BMI of living hunter-gatherers: J. R. Speakman, "Genetics of Obesity: Five Fundamental Problems with the Famine Hypothesis," in *Adipose Tissue and Adipokines in Health and Disease*, 2nd ed., ed. G. Fantuzzi and C. Braunschweig (Springer, 2014), 169–86.

none of them show: E. A. Brown, "Genetic Explorations of Recent Human Metabolic Adaptations: Hypotheses and Evidence," *Biological Reviews* 87, no. 4 (2012): 838–55, doi:10.1111/j.1469-185X.2012.00227.x; Q. Ayub et al., "Revisiting the Thrifty Gene Hypothesis via 65 Loci Associated with Susceptibility to Type 2 Diabetes," *American Journal of Human Genetics* 94, no. 2 (2014): 176–85, doi:10.1016/j.ajhg.2013.12.010.

quite the reverse: L. Segurel et al., "Positive Selection of Protective Variants for Type 2 Diabetes from the Neolithic Onward: A Case Study in Central Asia," *European Journal of Human Genetics* 21, no. 10 (2013): 1146–51, doi:10.1038/ejhg.2012.295.

According to Dr. Robert Lustig: R. H. Lustig, *Fat Chance: Beating the Odds against Sugar, Processed Food, Obesity, and Disease* (Penguin, 2012).

Fructose consumption: Ibid., 21.

caloric intake and expenditure: H. Pontzer et al., "Constrained Total Energy Expenditure and Metabolic Adaptation to Physical Activity in Adult Humans," *Current Biology* 26, no. 3 (February 8, 2016): 410–17, http://dx.doi.org/10.1016/j.cub.2015.12.046.

psychologists have discovered influence us: C. Spence and B. Piqueras-Fiszman, *The Perfect Meal: The Multisensory Science of Food and Dining* (Wiley Blackwell, 2014).

A study of obese patients: R. H. Lustig et al., "Isocaloric Fructose Restriction and Metabolic Improvement in Children with Obesity and Metabolic Syndrome," *Obesity* 24, no. 2 (February 2016), doi:10.1002/oby.21371.

calls fructose a toxin: R. H. Lustig et al., "The Toxic Truth about Sugar," *Nature* 482, no. 7383 (2012): 27, doi:10.1038/482027a.

"paleofantasy": M. Zuk, *Paleofantasy: What Evolution Really Tells Us about Sex, Diet, and How We Live* (Norton, 2013).


## CHAPTER ELEVEN

If it could be demonstrated: C. Darwin, *The Origin of Species by Means of Natural Selection* (reprint of the first edition; Penguin, 1859).

Is it conceivable that: quoted in O. T. Oftedal, "The Mammary Gland and Its Origin during Synapsid Evolution," *Journal of Mammary Gland Biology and Neoplasia* 7, no. 3 (2002).

glands that produce milk: C. M. Lefevre et al., "Evolution of Lactation: Ancient Origin and Extreme Adaptations of the Lactation System," *Annual Review of Genomics and Human Genetics* 11 (2010): 219–38, doi:10.1146/annurev-genom-082509-141806; O. T. Oftedal and D. Dhouailly, "Evo-Devo of the Mammary Gland," *Journal of Mammary Gland Biology and Neoplasia* 18, no. 2 (2013): 105–20, doi:10.1007/s10911-013-9290-8.

before the first mammals: O. T. Oftedal, "The Evolution of Milk Secretion and Its Ancient Origins," *Animal* 6, no. 3 (2012): 355–68, doi:10.1017/s1751731111001935.

adaptive function for both mother and infant: C. Holt and J. A. Carver, "Darwinian Transformation of a 'Scarcely Nutritious Fluid' into Milk," *Journal of Evolutionary Biology* 25, no. 7 (2012): 1253–63, doi:10.1111/j.1420-9101.2012.02509.x.

all over southwest Asia: R. P. Evershed et al., "Earliest Date for Milk Use in the Near East and Southeastern Europe Linked to Cattle Herding," *Nature* 455, no. 7212 (2008): 528–31, doi:10.1038/nature07180.

curd cheese: M. Salque et al., "Earliest Evidence for Cheese Making in the Sixth Millennium BC in Northern Europe," *Nature* 493, no. 7433 (2013): 522–25, doi:10.1038/nature11698.

earliest Neolithic farmers: J. Burger et al., "Absence of the Lactase-Persistence-Associated Allele in Early Neolithic Europeans," *Proceedings of the National Academy of Sciences of the United States of America* 104, no. 10 (2007): 3736–41, doi:10.1073/pnas.0607187104.

Caucasus Mountains: Y. Itan et al., "The Origins of Lactase Persistence in Europe," *PLOS Computational Biology* 5, no. 8 (2009), doi:10.1371/journal.pcbi.1000491.

lactase persistence alleles: A. Curry, "The Milk Revolution," *Nature* 500 (2013): 20–22.

calcium: O. O. Sverrisdottir et al., "Direct Estimates of Natural Selection in Iberia Indicate Calcium Absorption Was Not the Only Driver of Lactase Persistence in Europe," *Molecular Biology and Evolution* 31, no. 4 (2014): 975–83, doi:10.1093/molbev/msu049.

Saudi Arabia: N. S. Enattah et al., "Independent Introduction of Two Lactase-Persistence Alleles into Human Populations Reflects Different History of Adaptation to Milk

Culture," *American Journal of Human Genetics* 82, no. 1 (2008): 57–72, doi:10.1016/j .ajhg.2007.09.012.

survey of Irish cheeses: L. Quigley, "High-Throughput Sequencing for Detection of Sub- populations of Bacteria Not Previously Associated with Artisanal Cheeses," *Applied and Environmental Microbiology* 78 (2012): 5717–23.

marine environments: B. E. Wolfe et al., "Cheese Rind Communities Provide Tractable Systems for In Situ and In Vitro Studies of Microbial Diversity," *Cell* 158, no. 2 (2014): 422–33, doi:10.1016/j.cell.2014.05.041.

shares with the nasty *Streptococcus*: Y. J. Goh et al., "Specialized Adaptation of a Lactic Acid Bacterium to the Milk Environment: The Comparative Genomics of *Streptococcus thermophilus* LMD-9," *Microbial Cell Factories* 10 (2011), doi:10.1186/1475-2859 -10-s1-s22.

pages of a book: J. Ropars et al., "A Taxonomic and Ecological Overview of Cheese Fungi," *International Journal of Food Microbiology* 155, no. 3 (2012): 199–210, doi:10.1016/j .ijfoodmicro.2012.02.005.

comparison of the genetics: G. Gillot et al., "Insights into *Penicillium roqueforti* Mor- phological and Genetic Diversity," *PLOS ONE* 10, no. 6 (2015), doi:10.1371/journal .pone.0129849.

hundreds of bacterial species: L. Quigley et al., "The Complex Microbiota of Raw Milk," *FEMS Microbiology Reviews* 37 (2013): 664–98, doi:10.1111/1574-6976.12030.

microbes: T. P. Beresford et al., "Recent Advances in Cheese Microbiology," *Interna- tional Dairy Journal* 11 (2001): 259–74.

aroma of cheese: E. J. Smid and M. Kleerebezem, "Production of Aroma Compounds in Lactic Fermentations," *Annual Review of Food Science and Technology* 5, ed. M. P. Doyle and T. R. Klaenhammer (2014): 313–26.

wild-type *L. lactis*: D. Cavanagh et al., "From Field to Fermentation: The Origins of *Lac- tococcus lactis* and Its Domestication to the Dairy Environment," *Food Microbiology* 47 (2015): 45–61, doi:10.1016/j.fm.2014.11.001.

making the bacterial genes: H. Bachmann et al., "Microbial Domestication Signa- tures of *Lactococcus lactis* Can Be Reproduced by Experimental Evolution," *Genome Research* 22, no. 1 (2012): 115–24, doi:10.1101/gr.121285.111.

how the effects of disuse: Darwin, *The Origin of Species*, chap. 5.

propionic acid bacteria: E. J. Smid and C. Lacroix, "Microbe-Microbe Interactions in Mixed Culture Food Fermentations," *Current Opinion in Biotechnology* 24, no. 2 (2013): 148–54, doi:10.1016/j.copbio.2012.11.007.

cooperate in the fermentation of yogurt: K. Papadimitriou et al., "How Microbes Adapt to a Diversity of Food Niches," *Current Opinion in Food Science* 2 (2015): 29–35, doi:10.1016/j.cofs.2015.01.001.

bacteriocins: P. D. Cotter et al., "Bacteriocins: Developing Innate Immunity for Food," *Nature Reviews Microbiology* 3, no. 10 (2005): 777–88.

toxin that kills yeast: K. Cheeseman et al., "Multiple Recent Horizontal Transfers of a Large Genomic Region in Cheese Making Fungi," *Nature Communications* 5 (2014): 2876, doi:10.1038/ncomms3876.

## CHAPTER TWELVE

hundreds of kinds: N. A. Bokulich et al., "Microbial Biogeography of Wine Grapes Is Conditioned by Cultivar, Vintage, and Climate," *Proceedings of the National Academy of Science USA* (2013), doi:10.1073/pnas.1317377110.

*Dekkera, Pichia*, and *Kloeckera*: I. Tattersall and R. DeSalle, *A Natural History of Wine* (Yale University Press, 2015).

ancestors of modern brewer's yeast: A. Hagman et al., "Yeast 'Make-Accumulate-Consume' Life Strategy Evolved as a Multi-Step Process That Predates the Whole Genome Duplication," *PLOS ONE* 8, no. 7 (2013), doi:10.1371/journal.pone.0068734.

two *ADH* genes: J. M. Thomson et al., "Resurrecting Ancestral Alcohol Dehydrogenases from Yeast," *Nature Genetics* 37, no. 6 (2005): 630–35.

13 and 21 Ma: M. A. Carrigan et al., "Hominids Adapted to Metabolize Ethanol Long before Human-Directed Fermentation," *Proceedings of the National Academy of Sciences of the United States of America* 112, no. 2 (2015): 458–63, doi:10.1073/pnas.1404167111.

pages of this book: There are 380 amino acids in *ADH4*. http://www.uniprot.org/uniprot/P08319#sequences (accessed December 27, 2015).

increase by 40 times: N. J. Dominy, "Ferment in the Family Tree," *Proceedings of the National Academy of Sciences of the United States of America* 112, no. 2 (2015): 308–9, doi:10.1073/pnas.1421566112.

human taste for alcohol: R. Dudley, *The Drunken Monkey: Why We Drink and Abuse Alcohol* (University of California Press, 2014).

ADH1B: T. D. Hurley and H. J. Edenberg, "Genes Encoding Enzymes Involved in Ethanol Metabolism," *Alcohol Research: Current Reviews* 34, no. 3 (2012): 339–44.

much less likely: D. W. Li et al., "Strong Association of the Alcohol Dehydrogenase 1B Gene (*ADH1B*) with Alcohol Dependence and Alcohol-Induced Medical Diseases," *Biological Psychiatry* 70, no. 6 (2011): 504–12, doi:10.1016/j.biopsych.2011.02.024.

cardiovascular disease: M. V. Holmes et al., "Association between Alcohol and Cardiovascular Disease: Mendelian Randomisation Analysis Based on Individual Participant Data," *BMJ* 349 (2014), doi:10.1136/bmj.g4164.

Two such mutations: Hurley and Edenberg, "Genes Encoding Enzymes Involved in Ethanol Metabolism."

coprine: http://en.wikipedia.org/wiki/Coprinopsis_atramentaria#Toxicity (accessed December 30, 2015).

*Lactococcus chungangensis*: M. Konkit et al., "Alcohol Dehydrogenase Activity in *Lactococcus chungangensis*: Application in Cream Cheese to Moderate Alcohol Uptake," *Journal of Dairy Science* 98, no. 9 (2015): 5974–82, doi:10.3168/jds.2015-9697.

to brew beer: B. Hayden et al., "What Was Brewing in the Natufian? An Archaeological Assessment of Brewing Technology in the Epipaleolithic," *Journal of Archaeological Method and Theory* 20, no. 1 (2013): 102–50, doi:10.1007/s10816-011-9127-y.

Neolithic village of Jiahu: P. E. McGovern et al., "Fermented Beverages of Pre- and Proto-Historic China," *Proceedings of the National Academy of Sciences of the United States of America* 101, no. 51 (2004): 17593–98.

wild plants: P. This et al., "Historical Origins and Genetic Diversity of Wine Grapes," *Trends in Genetics* 22, no. 9 (2006): 511–19, doi:10.1016/j.tig.2006.07.008.

earliest archaeological evidence: P. E. McGovern et al., "Neolithic Resinated Wine," *Nature* 381, no. 6582 (1996): 480–81, doi:10.1038/381480a0.

village of Areni in Armenia: H. Barnard et al., "Chemical Evidence for Wine Production around 4000 BCE in the Late Chalcolithic Near Eastern Highlands," *Journal of Archaeological Science* 38, no. 5 (2011): 977–84, doi:10.1016/j.jas.2010.11.012.

Ian Tattersall and Rob DeSalle: Tattersall and DeSalle, *A Natural History of Wine*.

genetics does corroborate: S. Myles et al., "Genetic Structure and Domestication History of the Grape," *Proceedings of the National Academy of Sciences of the United States of America* 108, no. 9 (2011): 3530–35, doi:10.1073/pnas.1009363108.

domesticated in the western Mediterranean: R. Arroyo-Garcia et al., "Multiple Origins of Cultivated Grapevine (*Vitis vinifera* L. ssp. *sativa*) Based on Chloroplast DNA Polymorphisms," *Molecular Ecology* 15, no. 12 (2006): 3707–14, doi:10.1111/j.1365-294X.2006.03049.x.

the Georgians: S. Imazio et al., "From the Cradle of Grapevine Domestication: Molecular Overview and Description of Georgian Grapevine (*Vitis vinifera* L.) Germplasm," *Tree Genetics and Genomes* 9, no. 3 (2013): 641–58, doi:10.1007/s11295-013-0597-9.

Santiago de Compostela: J. C. Santana et al., "Genetic Structure, Origins, and Relationships of Grapevine Cultivars from the Castilian Plateau of Spain," *American Journal of Enology and Viticulture* 61, no. 2 (2010): 214–24.

mutations in grape clones: G. Carrier et al., "Transposable Elements Are a Major Cause of Somatic Polymorphism in *Vitis vinifera L*," *PLOS ONE* 7, no. 3 (2012), doi:10.1371/journal.pone.0032973.

Transposable elements: O. Jaillon et al., "The Grapevine Genome Sequence Suggests Ancestral Hexaploidization in Major Angiosperm Phyla," *Nature* 449, no. 7161 (2007): 463–67, doi:10.1038/nature06148.

Pinot Blanc, Pinot Gris: F. Pelsy et al., "Chromosome Replacement and Deletion Lead to Clonal Polymorphism of Berry Color in Grapevine," *PLOS Genetics* 11, no. 4 (2015): e1005081, doi:10.1371/journal.pgen.1005081.

clones producing white grapes: S. Kobayashi et al., "Retrotransposon-Induced Mutations in Grape Skin Color," *Science* 304, no. 5673 (2004): 982, doi:10.1126/science.1095011.

anthocyanin-promoting gene: A. Fournier-Level et al., "Evolution of the *VvMybA* Gene Family, the Major Determinant of Berry Colour in Cultivated Grapevine (*Vitis vinifera* L.)," *Heredity* 104, no. 4 (2010): 351–62, doi:10.1038/hdy.2009.148.

phylloxera: C. Campbell, *The Botanist and the Vintner* (Algonquin Books, 2004).

wild grape from Texas: Ibid.

Concorde grape: J. Granett et al., "Biology and Management of Grape Phylloxera," *Annual Review of Entomology* 46 (2001): 387–412, doi:10.1146/annurev.ento.46.1.387.

survived in China: X. M. Zhong et al., " 'Cabernet Gernischt' Is Most Likely to Be 'Carmenère,' " *Vitis* 51, no. 3 (2012).

local drinks: J. L. Legras et al., "Bread, Beer and Wine: *Saccharomyces cerevisiae* Diversity Reflects Human History," *Molecular Ecology* 16, no. 10 (2007): 2091–102, doi:10.1111/j.1365-294X.2007.03266.x; G. Liti et al., "Population Genomics of Domestic and Wild Yeasts," *Nature* 458, no. 7236 (2009): 337–41, doi:10.1038/nature07743.

oak trees: K. E. Hyma and J. C. Fay, "Mixing of Vineyard and Oak-Tree Ecotypes of *Saccharomyces cerevisiae* in North American Vineyards," *Molecular Ecology* 22, no. 11 (2013): 2917–30, doi:10.1111/mec.12155.

Hornets: I. Stefanini et al., "Role of Social Wasps in *Saccharomyces cerevisiae* Ecology and Evolution," *Proceedings of the National Academy of Sciences of the United States of America* 109, no. 33 (2012): 13398, doi:10.1073/pnas.1208362109.

brewmaster's beard: http://www.rogue.com/rogue_beer/beard-beer/ (accessed January 6, 2016).

39 genes: S. Marsit and S. Dequin, "Diversity and Adaptive Evolution of Saccharomyces Wine Yeast: A Review," *FEMS Yeast Research* 15, no. 7 (2015), doi:10.1093/femsyr/fov067.

flor yeast: H. Alexandre, "Flor Yeasts of *Saccharomyces cerevisiae*—Their Ecology, Genetics and Metabolism," *International Journal of Food Microbiology* 167, no. 2 (2013): 269–75, doi:10.1016/j.ijfoodmicro.2013.08.021.

*Saccharomyces carlsbergensis*: J. Wendland, "Lager Yeast Comes of Age," Eukaryotic Cell 13, no. 10 (2014): 1256–65, doi:10.1128/EC.00134-14.

## CHAPTER THIRTEEN

five days of feasting: J. McCann, *Stirring the Pot: A History of African Cuisine* (C. Hurst, 2010).

Ethiopia has more livestock: http://www.wolframalpha.com/input/?i=cattle+per+capita +in+African+countries (accessed January 29, 2016).

special dishes: McCann, *Stirring the Pot*, 74.

drought and the spread of rinderpest: P. Webb and J. Von Braun, *Famine and Food Security in Ethiopia: Lessons for Africa* (John Wiley & Sons Canada, 1994).

killing between 600,000 and 1 million: S. Devereux, *Famine in the Twentieth Century* (Institute of Development Studies, 2000).

A third of Ethiopian households: Webb and Von Braun, *Famine and Food Security in Ethiopia*.

$150 million: http://news.bbc.co.uk/1/hi/world/africa/703958.stm (accessed January 17, 2016).

*The Descent of Man*: C. Darwin, *The Descent of Man, and Selection in Relation to Sex* (J. Murray, 1901).

eight cousins: M. Kohn, *A Reason for Everything* (Faber & Faber, 2004), 281.

Haldane: R. Clark, *J. B. S.: The Life and Work of J. B. S. Haldane* (Bloomsbury, 2011).

Comparative studies: R. Kurzban et al., "The Evolution of Altruism in Humans," *Annual Review of Psychology* 66, ed. S. T. Fiske (2015): 575–99.

Richard Dawkins: Kohn, *A Reason for Everything*, 272.

Among primates: A. V. Jaeggi and C. P. Van Schaik, "The Evolution of Food Sharing in Primates," *Behavioral Ecology and Sociobiology* 65, no. 11 (2011): 2125–40, doi:10.1007/s00265-011-1221-3.

Cicero: M. Ridley, *The Origins of Virtue* (Viking, 1996).

Dear Septicius Clarus: A. Dalby and S. Grainger, *The Classical Cookbook* (British Museum Press, 1996), 100.

how hunter-gatherers share food: M. Gurven, "To Give and to Give Not: The Behavioral Ecology of Human Food Transfers," *Behavioral and Brain Sciences* 27, no. 4 (2004): 543–83; A. V. Jaeggi and M. Gurven, "Reciprocity Explains Food Sharing in Humans and Other Primates Independent of Kin Selection and Tolerated Scrounging: A Phylogenetic Meta-Analysis," *Proceedings of the Royal Society of London: Series B, Biological Sciences* 280, no. 1768 (2013), doi:10.1098/rspb.2013.1615.

animal societies: T. Clutton-Brock, "Cooperation between Non-Kin in Animal Societies," *Nature* 461, no. 7269 (2009): 51–57.

the best strategy: M. Tomasello et al., "Two Key Steps in the Evolution of Human Cooperation: The Interdependence Hypothesis," *Current Anthropology* 53, no. 6 (2012): 673–92, doi:10.1086/668207.

shared it under duress: I. C. Gilby, "Meat Sharing among the Gombe Chimpanzees: Harassment and Reciprocal Exchange," *Animal Behaviour* 71 (2006): 953–63, doi:10 .1016/j.anbehav.2005.09.009.

Gombe: Ibid.

oxytocin: R. M. Wittig et al., "Food Sharing Is Linked to Urinary Oxytocin Levels and Bonding in Related and Unrelated Wild Chimpanzees," *Proceedings of the Royal Society of London: Series B, Biological Sciences* 281, no. 1778 (2014), doi:10.1098 /rspb.2013.3096.

children happily share food: Tomasello et al., "Two Key Steps in the Evolution of Human Cooperation."

nor quite possibly do they care: J. M. Engelmann et al., "The Effects of Being Watched on Resource Acquisition in Chimpanzees and Human Children," *Animal Cognition* 19, no. 1 (2016): 147–51, doi:10.1007/s10071-015-0920-y.

photograph of a pair of eyes: M. Bateson et al., "Cues of Being Watched Enhance Cooperation in a Real-World Setting," *Biology Letters* 2, no. 3 (2006): 412–14, doi:10.1098 /rsbl.2006.0509.

Othello: W. Shakespeare, *Othello*, in *Complete Works of William Shakespeare RSC Edition*, ed. J. Bate and E. Rasmussen (Macmillan, 2006), 3.3.

*Boar á la Troyenne*: A. Soyer, *The Pantropheon: Or, a History of Food and Its Preparation in Ancient Times* (Paddington Press, 1977).

engastration escalation: www.dailymail.co.uk/news/article-502605/It-serves-125-takes -hours-cook-stuffed-12-different-birds-really-IS-Christmas-dinner.html (accessed February 9, 2016).

potlatch feast: Ridley, *The Origins of Virtue*.

New Guinea: B. Hayden, *The Power of Feasts* (Cambridge University Press, 2014).

## CHAPTER FOURTEEN

climate change: A. J. Challinor et al., "A Meta-Analysis of Crop Yield under Climate Change and Adaptation," *Nature Climate Change* 4, no. 4 (2014): 287–91, doi:10.1038 /nclimate2153.

adapt food production: B. McKersie, "Planning for Food Security in a Changing Climate," *Journal of Experimental Botany* 66, no. 12 (2015): 3435–50, doi:10.1093/jxb /eru547.

No hunter of the Age of Fable: A. D. Hope, "Conversations with Calliope," in *Collected Poems, 1930–1970* (Angus and Robertson, 1972), http://www.poetrylibrary.edu.au /poets/hope-a-d/conversation-with-calliope-0146087 (accessed February 20, 2016).

Two key books: P. R. Ehrlich, *The Population Bomb* (Ballantine, 1968); D. H. Meadows, *The Limits to Growth: A Report for the Club of Rome's Project on the Predicament of Mankind* (Earth Island Ltd., 1972).

food supply: L. T. Evans, *Feeding the Ten Billion: Plants and Population Growth* (Cambridge University Press, 1998).

natural habitat: J. R. Stevenson et al., "Green Revolution Research Saved an Estimated 18 to 27 Million Hectares from Being Brought into Agricultural Production," *Proceedings of the National Academy of Sciences of the United States* 110, no. 21 (2013): 8363.

Borlaug: N. Borlaug, "Norman Borlaug—Nobel Lecture: The Green Revolution, Peace, and Humanity," 1970, http://www.nobelprize.org/nobel_prizes/peace/laureates/1970 /borlaug-lecture.html (accessed February 20, 2016).

raising average yields: Evans, *Feeding the Ten Billion*.

current trends: D. K. Ray et al., "Yield Trends Are Insufficient to Double Global Crop Production by 2050," *PLOS ONE* 8, no. 6 (2013), doi:10.1371/journal.pone.0066428.

reducing food waste: M. Kummu et al., "Lost Food, Wasted Resources: Global Food Supply Chain Losses and Their Impacts on Freshwater, Cropland, and Fertiliser Use," *Science of the Total Environment* 438 (2012): 477–89, doi:10.1016/j.scitotenv .2012.08.092.

eat less meat: V. Smil, *Should We Eat Meat? Evolution and Consequences of Modern Carnivory* (Wiley-Blackwell, 2013).

mutations responsible: A. Sasaki et al., "Green Revolution: A Mutant Gibberellin-Synthesis Gene in Rice—New Insight into the Rice Variant That Helped to Avert Famine over Thirty Years Ago," *Nature* 416, no. 6882 (2002): 701–2, doi:10.1038/416701a.

salt tolerance: R. Munns et al., "Wheat Grain Yield on Saline Soils Is Improved by an Ancestral Na$^+$ Transporter Gene," *Nature Biotechnology* 30, no. 4 (2012): 360–64, doi:10.1038/nbt.2120.

mechanism of photosynthesis: S. P. Long et al., "Meeting the Global Food Demand of the Future by Engineering Crop Photosynthesis and Yield Potential," *Cell* 161, no. 1 (2015): 56–66, doi:10.1016/j.cell.2015.03.019; J. Kromdijk et al., "Improving Photosynthesis and Crop Productivity by Accelerating Recovery from Photoprotection," *Science* 354, no. 6314 (2016): 857–61, doi:10.1126/science.aai8878.

Romania: P. Ronald and R. W. Adamchak, *Tomorrow's Table: Organic Farming, Genetics, and the Future of Food*, 2nd ed. (Oxford University Press, 2017).

a survey: C. Funk and L. Rainie, "Public Opinion about Food," in *Americans, Politics and Science Issues* (Pew Research Center, 2015).

misled: W. Saletan, "Unhealthy Fixation," *Slate.com*, July 15, 2015, http://www.slate.com/articles/health_and_science/science/2015/07/are_gmos_safe_yes_the_case_against_them_is_full_of_fraud_lies_and_errors.html (accessed August 19, 2016).

safety of GM crops: A. Nicolia et al., "An Overview of the Last 10 Years of Genetically Engineered Crop Safety Research," *Critical Reviews in Biotechnology* 34, no. 1 (2014): 77–88, doi:10.3109/07388551.2013.823595.

now argue instead: H. van Bekkem and W. Pelegrina, "Food Security Can't Wait for GE's Empty Promises," June 30, 2016, http://www.greenpeace.org/international/en/news/Blogs/makingwaves/food-security-GE-empty-promises/blog/56913/ (accessed August 20, 2016).

benefits of GM technology: National Academies of Sciences Engineering and Medicine, *Genetically Engineered Crops: Experiences and Prospects* (National Academies Press, 2016), doi:10.17226/23395.

Papaya modified: D. Gonsalves, "Control of Papaya Ringspot Virus in Papaya: A Case Study," *Annual Review of Phytopathology* 36 (1998): 415–37, doi:10.1146/annurev.phyto.36.1.415.

Thailand in 2004: S. N. Davidson, "Forbidden Fruit: Transgenic Papaya in Thailand," *Plant Physiology* 147, no. 2 (2008): 487–93, doi:10.1104/pp.108.116913.

golden rice: Saletan, "Unhealthy Fixation."

denied access to GM varieties: R. L. Paarlberg, *Starved for Science: How Biotechnology Is Being Kept Out of Africa* (Harvard University Press, 2008).

GM eggplant: E. Hallerman and E. Grabau, "Crop Biotechnology: A Pivotal Moment for Global Acceptance," *Food and Energy Security* 5, no. 1 (2016): 3–17, doi:10.1002/fes3.76.

sustainable agriculture: Ronald and Adamchak, *Tomorrow's Table*.

Mark Lynas: M. Lynas, "How I Got Converted to GMO Food," *New York Times*, April 24, 2015.

GMOs are impossible to define: N. Johnson, "It's Practically Impossible to Define 'GMOs,'" December 21, 2015, https://grist.org/food/mind-bomb-its-practically-impossible-to-define-gmos/ (accessed March 20, 2016).

*Rhizobium radiobacter*: M. Van Montagu, "It Is a Long Way to GM Agriculture," *Annual Review of Plant Biology* 62 (2011): 1–23, doi:10.1146/annurev-arplant-042110-103906.

Genome of the domesticated sweet potato: T. Kyndt et al., "The Genome of Cultivated Sweet Potato Contains *Agrobacterium* T-DNAs with Expressed Genes: An Example of a Naturally Transgenic Food Crop," *Proceedings of the National Academy of Sciences* 112, no. 18 (2015): 5844–49, doi:10.1073/pnas.1419685112.

CRISPR-Cas9: J. A. Doudna and E. Charpentier, "The New Frontier of Genome Engineering with CRISPR-Cas9," *Science* 346, no. 6213 (2014), doi:10.1126/science.1258096.

wheat susceptible to mildew: S. Huang et al., "A Proposed Regulatory Framework for Genome-Edited Crops," *Nature Genetics* 48, no. 2 (2016): 109–11, doi:10.1038/ng.3484, http://www.nature.com/ng/journal/v48/n2/abs/ng.3484.html#supplementary-information (accessed March 12, 2014).

Dietary studies confirm: C. T. McEvoy et al., "Vegetarian Diets, Low-Meat Diets and Health: A Review," *Public Health Nutrition* 15, no. 12 (2012): 2287–94, doi:10.1017/s1368980012000936.

recipe book: D. Bateson and W. Janeway, *Mrs. Charles Darwin's Recipe Book: Revived and Illustrated* (Glitterati, 2008).

# Index

Page numbers in italics refer to maps.